COLD FUSION

The Making
of a Scientific
Controversy

F. DAVID PEAT

CB
CONTEMPORARY
BOOKS
CHICAGO

Library of Congress Cataloging in Publication Data

Peat, F. David, 1938–
    Cold fusion.

        1. Controlled fusion.   2. Controlled fusion—
Research—Utah.   I. Title.
QC791.735.P43   1989   539.7'64   89-25160
ISBN 0-8092-4243-5 (cloth)
        0-8092-4085-8 (paper)

Published by Contemporary Books, Inc.
180 North Michigan Avenue, Chicago, Illinois 60601
Manufactured in the United States of America
International Standard Book Number: 0-8092-4243-5 (cloth)
                                                            0-8092-4085-8 (paper)

Contents

Acknowledgments

I would like to thank my agent, Adele Leone, and Contemporary Books for moving so rapidly on this project. Thanks are also due to the following:

At Brigham Young University: Bart Czirr, Paul Decker, Gary Jensen, Steven Jones, and Paul Palmer, and special thanks to Nancy Perkins for all her help.

At the University of Utah: Jim Brophy, Hugo Rossi, Jack Simons, Rick Steiner, and Cheves Walling.

Chandre Dharma-wardana of the National Research Council of Canada and Arthur Cordell of the Canadian Department of Communications for helpful discussions.

Jake Morrison, Bill Page, and Craig Taylor for introducing me to computer networks.

Chapter 1
The Utah Bombshell

On March 23, 1989, two chemists called a press conference at the University of Utah in Salt Lake City. The announcement they made was staggering. In the words of one scientist, "It is as important as the discovery of fire."

"Simple experiment results in sustained N-fusion at room temperature for first time. Breakthrough process has potential to provide inexhaustible source of energy," read the headlines in the University of Utah's press release. Martin Fleischmann and B. Stanley Pons claimed to have done nothing less than create nuclear fusion in a test tube. Using apparatus that can be found in a school laboratory, they had harnessed the power of the sun. For over 100 hours, Fleischmann and Pons had created pure energy in a tiny glass jar. The fuel they were using was as abundant as seawater.

Controlled nuclear fusion has been a dream for decades. If fusion power could be harnessed, unlimited energy would be available to the whole world. But everyone believed that controlled fusion power lay decades away and demanded massive equipment, experiments on an international scale, and funds of tens of billions of dollars a year. Now two chemists had done the whole thing in a small basement laboratory.

Within hours of the Utah announcement, the entire scientific community was in a state of shock. Physicists, chemists, and engineers talked to each other over computer networks. Fax machines spewed out letters and page upon page of calculations. The wire services buzzed. Scientists from one of the United States' major laboratories sat glued to a television set, trying to learn more about Fleischmann and Pons's discovery from a TV news item.

Science had never been done this way before. Scientific conferences and publications tend to be planned long in advance, so over the following days, scientists could only rely upon news reports for current information. Graduate students worked around the clock to duplicate the Fleischmann and Pons experiment. Distinguished scientists dropped their research projects and began to build fusion cells. In Canada, Ontario Hydro, the world's major supplier of heavy water—the fuel that had been used by Fleischmann and Pons in their cell—was inundated by telephone calls from laboratories, companies, and even schoolchildren who wanted to try the fusion experiment.

Everyone was getting "fusion fever." Things were becoming so frantic that one Nobel Prize winner cautioned that science was in danger of falling into chaos because the "due process" of scientific investigation—which requires a methodical peer review of any new discovery—had been bypassed.

Science was very much in the news as, with each edition, major newspapers carried updates of the fusion story. As the next weeks went by, one laboratory would announce a confirmation of the Fleischmann and Pons experiment, while another would claim that nothing spectacular had happened when it had attempted to replicate the experiment. One theoretician would prove that the whole thing was a storm in a teacup, simply the result of a well-known chemical reaction that had no potential as a new energy source. Another would offer proof of novel nuclear reactions producing immense energy with very little radiation.

In short, no one really knew what to think. Except, that is, for the financial speculators who were prepared to make a killing on the power of rumor alone. Futures on palladium (the metal that had been used in the Fleischmann and Pons experiment) rose to a five-year high before retreating.

In a new statement released on March 31, a few days after their original press conference, Pons and Fleischmann left no doubt of what they had achieved. The experiment had generated an abundant supply of energy, they said, that persisted for over 100 hours. All this using a fuel that is present in ordinary water. And this was only the beginning, for Fleischmann and Pons believed that the process could be scaled up to solve the energy problems of the human race for incalculable centuries to come.

Many experienced scientists, however, could not believe that so much energy could be produced in a test tube. Robert Cohen, director of UCLA's Institute of Plasma and Fusion Research, declared that if this energy was really being produced by conventional nuclear fusion, then the amount of radiation released in the process would have killed everyone in the laboratory.

The whole idea of Utah's cold fusion sounded crazy. It did not agree with what everyone took to be the laws of nuclear physics. "It's got to be wrong," said Stanley Luckhardt, a physicist with MIT's fusion group. "I'm afraid we'll look like idiots if we are seen trying this thing." Said that group's head, Ronald R. Parker, "My first reaction was that it was incredible. In fusion research there are always crackpot claims to produce fusion in a simple way. It always turns out that a little green man from Mars told them how to do it. When I heard this, I thought . . . here's another one, but for some reason the *Wall Street Journal* bit on this one."

A matter of days after the original announcement by Fleischmann and Pons, the cold-fusion plot took on a sudden twist. A second group announced that it had discovered room-temperature fusion; moreover, this bombshell came from a university only forty miles away, in Provo, Utah. The announcement from Brigham Young University was more guarded in its claims. To begin with, although BYU's scientists had not seen the same large amounts of heat being produced as had the University of Utah team, radioactive particles (neutrons) were certainly detected. Group leader Steven E. Jones said, "The discovery of cold nuclear fusion . . . opens the possibility, at least, of a new path to fusion energy."

In fact, it would later become known that the group from Brigham Young University, including a scientist from the University of

Arizona, had been working along the same track as the University of Utah team for several years. By now, rumors were beginning to circulate that the two groups already knew about each other's work, that a grant application made by Pons and Fleischmann had been sent to Steven Jones as an independent referee, and that an agreement had been made to announce cold fusion on the same day. Claims were also made that the two teams had been collaborating with each other, as well as, contrarily, that the two universities were involved in a heavy competition. The major question that people had begun to ask themselves was why Fleischmann and Pons had jumped the gun and gone to the press before Jones and his team from Brigham Young University. Was this a race for the Nobel Prize, or were some vital patents at stake?

Whatever the story, the news from Brigham Young University appeared to be an independent corroboration of Fleischmann and Pons's work. Now scientists were prepared to take the whole thing a bit more seriously. In the words of Carl Henning from the Lawrence Livermore National Laboratory, "While it's too early to say for sure, the general opinion is that there may be something in it."

Over the next few weeks, hopes were to rise and fall as evidence of new experiments from the United States, the Soviet Union, Italy, India, Czechoslovakia, and the United Kingdom began to come in over the wire services. At first some reports appeared to confirm the Fleischmann and Pons phenomenon; then others questioned that anything out of the ordinary was happening. By mid-April different laboratories were making, almost daily, conflicting claims. Was the promise of nuclear fusion slipping through our fingers? Or was it simply a matter of time before the final, crucial confirmation appeared?

FUSION POWER

What is nuclear fusion, and why should the University of Utah's announcement have sent such a frisson of excitement through the scientific and business communities? For tens of thousands of years, since the first protohumans appeared on earth, the long march of civilization has depended upon the availability of chemical energy. Humans have relied on chemical energy to provide them with warmth and light, to cook their food, and to make everything from cooking pots and bronze ax heads to high-definition television sets and next year's model car. Fusion power essentially multiplies the power available to the human race by a phenomenally large factor while avoiding many of the disadvantages associated with conventional power sources.

A familiar example of chemical energy is the hunter's fire. Sticks of wood burn in the hearth and give off heat and a little light. The chemical reaction taking place in the hunter's fire is not all that different from the burning of coal or oil in an electricity generating station. Essentially it is a chemical reaction between oxygen in the air and carbon compounds in the fuel. The result is the production of another chemical, carbon dioxide, along with the release of energy in the form of heat and light.

Figure 1-1

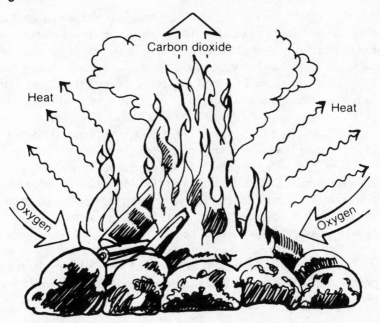

But there is a problem: the hunter must sleep lightly, for every few hours, new sticks of wood have to be thrown on the fire. Similarly the generating station must be constantly fed with many tons of coal. In your basement, oil is pumped from the reservoir tank into your furnace each time it switches on.

Figure 1–2

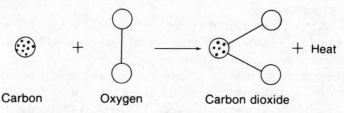

Carbon   Oxygen   Carbon dioxide

Obtaining energy from a chemical reaction requires a constant expenditure of fuel. Each day, the hunter has to spend time collecting wood and must have at times dreamed of getting more heat and light out of every armful of fuel.

Chemically speaking, it is a short step from the hunter's fire to the coal-fired generating station. In the latter case, coal is burned in a forced draft of air inside a furnace. Again the chemical reaction between the fuel and oxygen in the air creates carbon dioxide and heat. This heat is used to produce steam, which drives the steam turbines in the generating station. The furnace in your home, the engine in your car, the aircraft that takes you on a business trip, the rockets that carry a space shuttle into orbit all use chemical reactions to create their power—in every case fuel is being burned.

Even the energy expended in the simple act of walking has a chemical origin. Sugars derived from food are burned in the oxygen that is absorbed in the lungs and circulated in the blood. The energy released in this "burning" powers the body, and with each breath, carbon dioxide is released into the air.

Until the fifth decade of our own century, the history of the world's technology had been based upon the availability of chemical energy. Our whole concept of what was achievable in the world was limited by the sorts of energies released in a chemical reaction.

Figure 1-3

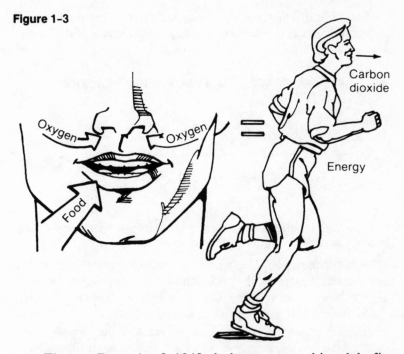

Then, on December 2, 1942, the human race achieved the first self-sustaining nuclear chain reaction. Working in a converted squash court at the University of Chicago, Enrico Fermi and his team built what they called an "atomic pile" and created the first nuclear reactor. In nuclear fission, the nuclei, or centers, of uranium atoms are encouraged to split; in this fission process, a large amount of energy is released—much more than in any chemical reaction. The key to a nuclear reactor is to allow this chain reaction to proceed in an orderly way so that useful energy can be extracted from a continuously running fission reaction.

Using nuclear fission in place of chemical reactions, scientists were able to extract much more energy from a given amount of fuel. The difference was impressive. To run a 1,000-megawatt generating station for a year requires twenty thousand railcars of coal. A similar volume of oil is needed for an oil-fired station. But when it comes to nuclear power, a single railcar of uranium will perform the same function. The difference is dramatic, and for a time it looked as if nuclear fission was the solution to many of the dreams of the human race.

Figure 1-4   A TYPICAL NUCLEAR FISSION REACTION

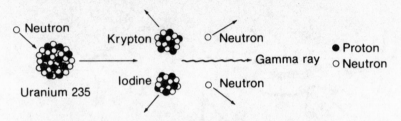

Uranium 235 absorbs a neutron and splits into Krypton and Iodine, releasing two more neutrons and a gamma ray.

Apart from generating its heat through nuclear fission, rather than chemical reactions, the nuclear power station operates in much the same way as an oil- or a coal-fired station. Heat is generated within the nuclear reactor and constantly removed by a cooling system. Depending upon the type of reactor, the closed-circuit cooling system uses water, gas, or even a liquid metal like sodium. Outside the reactor, this coolant gives up its heat through a system of heat exchangers. This heat is then used to produce steam that drives turbine generators, just as in a coal- or an oil-fired generating station. The steam that powers the generators never comes in direct contact with the nuclear reactor.

Not only does nuclear fission produce a phenomenally greater amount of energy for a given weight of fuel, but, environmentally speaking, it also has other advantages. Fossil fuels, like coal and oil, generally contain sulphur and nitrogen as well as other undesirable substances. In a year of running, a coal-fired generator produces 2,000 railcars of ash. Each day, 140 tons of sulphur and a similar amount of the oxides of nitrogen are released into the atmosphere. Both of these gases are the primary constituents of the acid rain that is destroying our lakes and forests, as well as the facades of our buildings.

Even more striking is the phenomenal amount of carbon dioxide

Figure 1-5   NUCLEAR FISSION POWER STATION

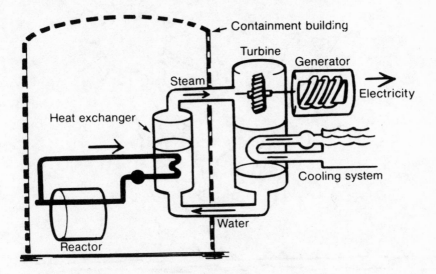

that is pumped into the atmosphere by a fossil-fired furnace. This carbon dioxide forms a blanket over the earth's atmosphere, preventing some of the sun's heat from radiating away into the cold of space. This phenomenon is called the greenhouse effect, and its direct consequence is a general warming of the earth. The implications of this climatic change are still being studied. Some climatologists predict disastrous effects—an increase in the size of deserts, depletion of valuable farmland, and melting of the ice caps along with a rise in the ocean levels that will threaten many cities located near the sea.

This combination of acid rain and the greenhouse effect—both the direct result of chemical reactions involving burning—has led many environmentalists to call for an end to the widespread use of fossil fuels (like oil and coal). A civilization that consumes—as we do—more and more energy from the burning of fossil fuels simply cannot go on, they warn. If the human race continues on this path, then the world will eventually become uninhabitable.

Nuclear fission does not generate carbon dioxide, nor does it produce the gases that lead to acid rain. Furthermore, the energy that

Figure 1-6

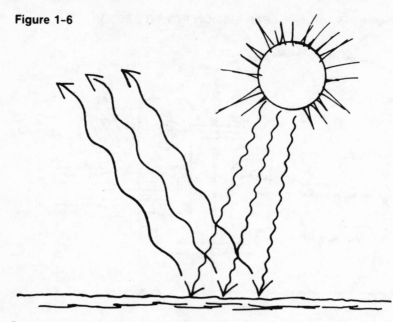

Some of the sun's energy is reflected from the earth.

can be obtained from a ton of uranium vastly exceeds that bound up in a ton of oil. Does nuclear fission therefore provide the answer to this energy crisis? Some think it does.

Others are concerned with the hazardous element involved in nuclear power. Radioactive waste is constantly being generated in the day-to-day operation of a nuclear power station. Inside each nuclear fuel rod, a number of highly radioactive products are building up as a result of the fission reactions within the reactor. At some point, new fuel rods must be inserted, and the spent, highly radioactive rods must be disposed of.

Although a high percentage of the radioactive isotopes within a spent fuel rod decay within decades, the lifetime of some of them is measured in centuries and tens of centuries. Will it really be possible to isolate these hazardous substances from the environment over such a long time span? Nuclear experts have proposed a number of solutions for what they feel is reliable long-term disposal. Some nuclear critics disagree.

Figure 1-7   THE GREENHOUSE EFFECT

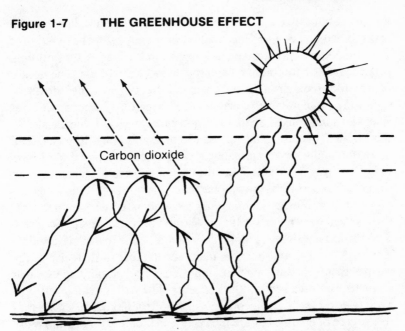

Carbon dioxide

Heat reflected from the earth is trapped by a layer of carbon dioxide and reflected back toward the earth.

An even greater problem associated with nuclear power is the danger of a reactor accident. Great care is taken in the design of a nuclear power station, and many safety measures and backup systems are incorporated. Nevertheless, reactor accidents have happened in a number of countries. The most serious of these took place at Chernobyl in the Soviet Union, releasing a considerable amount of radioactivity that drifted across Eastern Europe, eventually reaching Scandinavia and the United Kingdom. In many areas, food and water were contaminated. Radioactive isotopes, absorbed by plants which were then eaten by animals, found their way into the food chain and, for example, raised the level of radioactivity in lamb. The long-term effects of this disaster have yet to be assessed, but it is clear that the cost will be paid in human lives.

For those who believe that the potential hazards of nuclear fission outweigh its advantages, this technology is simply unacceptable. Many politicians and communities are now having second thoughts about their nuclear energy policies. It should, however, be realized that no large-scale technology is entirely safe. The bursting of a hydroelectric dam can cause death and widespread flooding. Propane tanks have fractured and spilled a curtain of fire. The release of hazardous materials from chemical plants can cause death or injury to a large number of people. The Union Carbide catastrophe at Bhopal, India, caused death or long-term injury on a wide scale. At some time in the future, the release of a biologically modified bacteria or virus could be even more devastating. Indeed, many thinkers are today having reservations about the unlimited pace of technology and the constant desire for growth and more energy.

But even if other technologies are associated with human risk, this by itself does not make the nuclear industry acceptable. So where does the answer lie? The most sensible solution is that societies must learn to conserve and become more economical in their use of the earth's resources. But even with improved conservation, societies will still need to find a new source of energy. The widespread burning of fossil fuels is clearly unacceptable. Nuclear fission provides an alternative source of energy, but are the risks acceptable? If more and more countries convert to nuclear power, then the risk of even one accident increases—and a serious accident can have global repercussions.

It is also important to realize that even nuclear power may not be a truly long-term solution to the world's energy needs. Although a much smaller amount of uranium is needed to produce a given amount of energy, the world's supply of uranium is finite. It is possible to buy more time by using what are called "breeder reactors," in which new fuel is actually produced during the burning of the old. But using breeder reactors means building factories in which the nuclear fuel is reprocessed, another hazardous process, and one that some countries find unacceptable.

If fossil fuels are causing environmental damage and the risks involved in the unlimited use of nuclear fission are not acceptable to the general public, then what is the answer?

For many scientists, the key to safe, clean, and unlimited energy is nuclear *fusion*—not fission. Nuclear fusion is the process that powers the sun and stars. It involves the fusion, or joining, of the nuclei of hydrogen atoms. The process is a little like the splitting of an amoeba—but played backward. Two hydrogen nuclei collide, fuse, and create a new nucleus. The result of nuclear fusion is inert helium gas. The by-product is an astronomical amount of energy. Running a 1,000-megawatt power station for a year requires twenty thousand railcars of coal; one powered by fusion needs only a pickup truck full of fuel. In fact, the consumption of less than a cupful of fusion fuel has the equivalent energy of three billion tons of chemical fuel.

Is the fuel for this energy of the stars a rare and exotic compound? No, it is found in rivers, lakes, and seawater. It is a naturally occurring form of water called heavy water which is present in a small proportion of water everywhere on earth. In fact, the heavy water in a gallon of ordinary seawater contains enough energy to drive a car from New York to San Francisco and back!

When all the world's coal, oil, and uranium have been used up, heavy water will still be around. It will be available as long as the rivers run and the rain falls. In fact, it has been estimated that the top ten feet of Lake Michigan contain enough heavy water to supply the energy needs of the United States for the next 15,000 years. The top few feet of water in the world's oceans contain enough energy to supply the world for 30 million years.

If fusion fuel is so abundant and its energy potential unlimited, then why is the world bothering about its energy supplies? Why is the price of oil so high, and why are fossil fuels being burned, causing acid rain and the buildup of carbon dioxide in the atmosphere? The reason is that the power of the sun has proved extraordinarily difficult to tame. Despite several decades of work by leading physicists around the world and the expenditure of many billions of dollars, progress in nuclear fusion has been very slow.

With the commissioning of each new experimental fusion machine, scientists creep closer to the break-even point at which more energy is produced than has to be expended in initiating the fusion reaction. But each step forward is very small, and enormous technical difficulties must be solved before nuclear fusion can become a commercial possibility. Some experts believe that this will not happen until around 2050, and that when it does, the machines will be so large and costly that the final venture can be done only on an international scale.

THE UTAH SOLUTION

This is why the Fleischmann and Pons result came as such a shock. It simply seemed impossible that two chemists had achieved the breakthrough that had occupied armies of physicists and engineers for decades. For most physicists, nuclear fusion meant temperatures of tens of millions of degrees, ultimately to be attained by the use of gigantic magnets or enormously powerful lasers. Yet Fleischmann and Pons claimed to have tamed the power of the sun in a test tube. Simply powering up a nuclear fusion machine requires the electrical energy of a small town, yet Fleischmann and Pons claimed to have done the whole thing using a car battery. While the machinery for one of the major fusion labs requires something the size of a factory, Fleischmann and Pons claimed to have achieved the reaction on a laboratory bench.

For some scientists, the Fleischmann and Pons announcement produced a sense of elation. Here was a totally new phenomenon, something profound and unexpected. Nature had revealed yet another card in its deck.

But for others, it was a matter of skepticism. "I don't believe nature would work this way," was one reaction. "The whole thing sounds like a free lunch, and the universe never does that sort of thing."

The events of the days and weeks that followed Fleischmann and Pons's announcement did nothing to allay this sense of confusion. Physicists were seeking definitive answers, awaiting some crucial experiment that would resolve the uncertainty one way or the other. But this simply did not come about.

Fleischmann and Pons claimed to have noted two major clues that indicated a fusion reaction was going on. First, an enormous amount of heat was generated in their apparatus—the sort of heat that could never be produced in a chemical reaction. Second, the sort of radiation that is associated with a nuclear reaction had been detected. Nuclear radiation is not associated with chemical reactions like burning.

The key question everyone was asking was whether this intense heat and nuclear radiation could be confirmed by an independent study. In the weeks that followed, some laboratories found heat but no radiation. Others detected radiation but no heat. Some, after very careful experiments, found neither.

Even the two Utah groups did not agree on what they were observing. While the laboratory at the University of Utah had detected so much heat that in one case the apparatus melted and disintegrated, those at Brigham Young University found the heat too minute to measure! What sort of a phenomenon was it that presented a totally different face to each experimental group? It was like the many-faced god Janus of Roman legend, something scientists found very hard to take. Science is supposed to be reproducible, and repeated experiments generally give basically similar results. But here was a world-shattering phenomenon that spoke in a very different language depending on whether you happened to interrogate it in Salt Lake City, Utah, or in Provo, Utah. What was going on?

THE SHOWDOWN

By April 26, Fleischmann and Pons were appearing before a congressional committee and asking for $25 million to help commercialize their findings. Jones, on the other hand, warned the members of Congress to be cautious about large-scale funding of a field that was still in its infancy. What were the legislators to believe? Already a number of confirmations had been announced, in the United States and abroad. But rumors were circulating that other groups had had negative results and even that Fleischmann and Pons's work had been discredited. No longer just a story about a possible major breakthrough, a whole human drama was emerging.

A turning point in the story came at the beginning of May, when two landmark scientific meetings were held: the spring meeting of the American Physical Society in Baltimore, and the annual meeting of the Electrochemical Society at San Diego. At 7:30 in the evening of May 1, the world's eyes were turned to the Baltimore Convention Center, where at a special session that went on past midnight, scientists from all over the world recounted their own independent results on cold fusion. The scientific world was about to pass judgment on Fleischmann, Pons, and Jones.

Chapter 2
The Energy of the Sun

The announcement that nuclear fusion had been created in a test tube was tantamount to claiming that Fleischmann and Pons had tamed the reactions that power the sun and stars. To understand what nuclear fusion means and the incredible size of the energies that are released, what better way to begin than with the power produced by the sun itself.

THE SUN

The sun is made of gas, mainly hydrogen and helium, and has been shining in the sky, giving out its energy, for around 5 billion years. Astronomers say that it will still be shining 5 billion years into the future! It is clear that chemical reactions cannot keep the sun going. The sun is over 100 times greater in diameter than the earth, and even if it were made entirely of coal, at its present rate of energy production it could not have lasted for even a millionth of its present age.

For the technically minded, the sun's output of energy is an amazing 400,000,000,000,000,000,000,000,000 watts. (An electrical kettle uses around 1,000 watts; a light bulb consumes

100 watts.) This output energy is equivalent to the power of 400,000,000,000,000 large nuclear power stations working night and day. And this has been going on for thousands of millions of years.

Here on earth, 93 million miles away from the sun, we receive only one-half of one-billionth of this energy. Yet this energy is enough to warm and light us. It is responsible for all the world's weather, for the evolution of life, and for the fossil fuels like wood, coal, and oil that we burn today. The energy falling on even a piece of land the size of a tabletop is measured in kilowatts. Every second of its existence, the sun is putting out an unbelievable amount of energy, and yet it is just one of 100 billion stars in our galaxy alone!

Figure 2-1   THE SUN

One of 100 billion stars in our galaxy

Diameter: 864,950 mi. (1,392,000 km)

Mass: 4×10^{30} lbs. (2×10^{30} kg.), that is, 333,000 times the mass of the earth

Distance: 93×10^6 mi. from earth

Temperature
outer surface: 5,500° C

inner core: over $10^{6°}$ C

Burns: 4×10^6 tons of hydrogen gas/sec

Energy output: 4×20^{26} W or 4×10^{22} ergs/sec (equivalent to 400,000,000,000,000 nuclear power stations working night and day)

The sun's energy is unbelievable. We know that such a continuous output of power could never be produced by a chemical burning. Even nuclear fission is out of the question. In the first years of this century, there was considerable speculation as to what mysterious process could be powering the sun and stars. In fact, as scientists were soon to learn, the sun's energy is the direct result of nuclear fusion.

Nuclear fusion is a very different process than the more familiar nuclear fission—the sort of power that is used in nuclear power stations here on earth. In nuclear fission the nuclei of uranium atoms are encouraged to split and, in splitting, release energy. In fusion, however, two nuclei, the hearts of hydrogen atoms, meet and fuse. The result is the release of a tremendous amount of energy, far more than is released in nuclear fission.

Fusion is the secret power of the sun. Over its lifetime, the sun fuses hydrogen nuclei to produce helium. It is exactly this process that the two groups in Utah were attempting to reproduce in a test tube!

Early on in the atomic age, scientists had cracked the code to the energy of the stars. They also knew about its implications. The surface of the sun shines at a mere 5,500 degrees Celsius. While this may be as hot as a furnace on earth, it is very cold when compared to the temperature of over 10 million degrees Celsius at the sun's core. For it is only at such very high temperatures that conventional nuclear fusion becomes possible.

In this fusion reaction, the sun uses up 4 million tons of hydrogen gas every second. But there is still more than enough hydrogen to keep it going for another 5 billion years.

The nuclear reaction that fuses hydrogen produces helium. Helium could therefore be thought of as the "ash" left over from this nuclear furnace.

At the time of its birth, the sun contained 75 percent hydrogen and 25 percent helium. Today it has burned up some of this hydrogen, so that it now contains 65 percent hydrogen and 35 percent helium. Over the next few billion years, additional hydrogen will be consumed, and more and more helium will accumulate.

Even then the sun will not die. As its reserves of hydrogen become depleted, the sun will begin to cool and to become what is known as a red giant.

At this stage, our red-giant sun will have expanded to swallow the inner planets, including our own earth. Viewed from Saturn, it will appear like a great ball in the sky—no longer white hot but glowing a dull red.

As this cooling continues, our red-giant sun will collapse in on itself. In this sudden collapse, the temperature of its core will suddenly shoot up even higher, to the point where helium nuclei themselves begin to fuse and give off energy. A totally new form of nuclear fusion begins, and our sun will be reborn as an intensely hot white dwarf. Thanks to nuclear fusion, the sun is literally able to rise out of its own ash of helium and begin a new life.

The actual fusion reaction that powers the sun takes place only within its hot central core. Half of the sun's mass is concentrated into only 1.5 percent of its total volume, but within this tiny core 99 percent of its energy is created. Despite its size, the sun is a relatively gaseous body: halfway to its center, its density is only comparable with that of water, far less dense than the earth, for example. But at its core, the density is twelve times that of lead. It is here, under conditions of high pressure and extremely high temperatures, that nuclear fusion takes place.

The key to the incredible energies released through nuclear fusion lies in Einstein's famous equation, $E = mc^2$. In 1905 the young Einstein published his ground-breaking paper on the theory of relativity. Almost as an afterthought, he tagged on a separate three-page paper after his relativity publication. Its title is something of a mouthful: "Does the Inertia of a Body depend upon its Energy-Content?" Based on his previous paper, Einstein threw out a handful of deceptively simple equations and deduced that the mass of a substance is a measure of the energy it contains. In other words, if some of this mass were to disappear, then energy would be released.

How much energy? Einstein produced the formula: $E = mc^2$. The energy released (E) is equal to the mass (m) that disappears multiplied by the square of the speed of light (c). The speed of light is a very large number, and this means that a small amount of mass is equivalent to a very large amount of energy. The atomic energy contained in the mass of a chocolate bar is equivalent to the chemical energy released in the burning of several billion tons of coal!

The disappearance of mass, along with the creation of energy, is the clue to nuclear fusion. In the units used by atomic physicists, the mass of a hydrogen nucleus is given as 1.0078. A hydrogen nucleus contains one nuclear particle (a proton), while a helium nucleus contains four (two protons and two neutrons). It turns out that two acts of fusion, involving four hydrogen nuclei, are therefore required to produce a helium nucleus. The actual details of nuclear fusion are explored in greater detail in the next chapter.

Since four nuclei are involved, simple arithmetic gives a total mass of 4.0312 as the result. But it turns out that the mass of a helium nucleus is only 4.0030. This means that, in the fusion process, a mass of 0.0282 has disappeared. It is this disappearing mass that is converted directly into pure energy. Each time hydrogen nuclei fuse to produce helium, a tiny fraction of the sun's mass disappears in the form of heat and light. The sun is constantly losing mass, and it is this loss of mass that causes the trees to grow, the wind to blow, and our bodies to tan at the seaside.

Figure 2–2

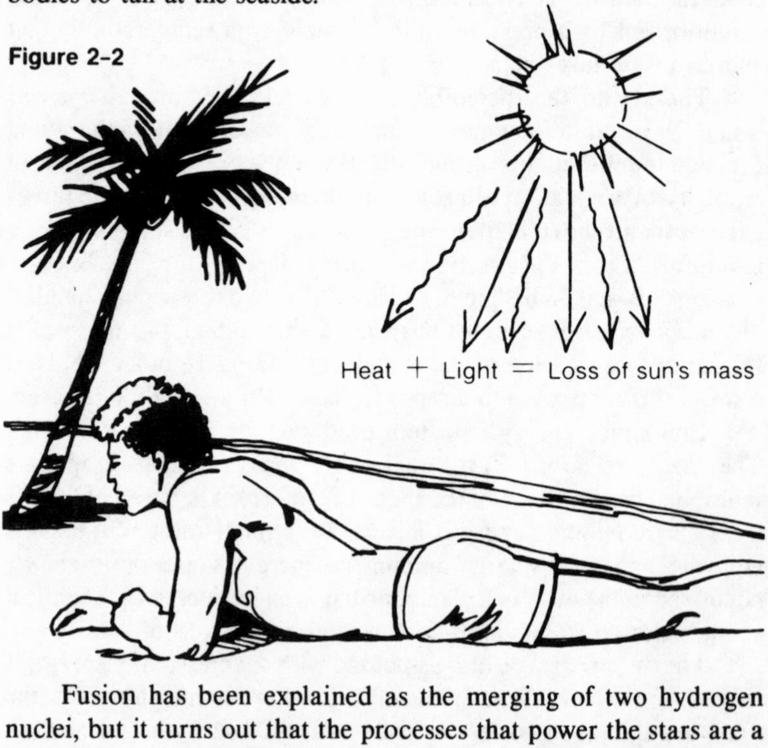

Heat $+$ Light $=$ Loss of sun's mass

Fusion has been explained as the merging of two hydrogen nuclei, but it turns out that the processes that power the stars are a little more complicated. The fusion of two hydrogen nuclei by themselves cannot directly lead to the creation of a helium nucleus.

Rather, the fusion process must proceed in a series of steps, each associated with a release of mass-energy. Scientists have carefully detailed the various fusion processes involved, and their calculations agree with the energy that is being radiated from the sun. The exact series of cycles involved depends upon the size of the star and the temperature of its interior.

THE CARBON CYCLE

Careful calculations showed scientists that the simple fusion of pairs of hydrogen nuclei (protons) would be too slow to account for the energy released within the sun's core. Although the protons are able to approach close enough, the actual fusion reaction takes too long.

While some stars do generate their energy through direct fusion of protons, the sun and other upper-main-sequence stars use carbon nuclei to speed up the process. This carbon cycle proceeds in the following way:

$$^{12}C + p = {}^{13}N$$
$$^{13}N = {}^{13}C$$
$$^{13}C + p = {}^{14}N$$
$$^{14}N + p = {}^{15}O$$
$$^{15}O = {}^{15}N$$
$$^{15}N + p = {}^{12}C + He$$

Note that in this sequence of reactions, protons (hydrogen nuclei) are constantly being "burned" to produce helium. The carbon 12 nucleus is released at the end of the reaction and is therefore free to catalyze more reactions.

The "energy barriers" to this carbon cycle are even higher than in the case of proton-proton fusion. However, once the nuclei have penetrated this barrier, the actual fusion reaction proceeds rapidly.

THE HYDROGEN BOMB

The same disappearance of mass during nuclear fusion is responsible for the intense heat and energy released in a hydrogen bomb. In fact, as soon as the building of the first A-bomb (which relies on the release of energy from nuclear fission) became a practical possibility, scientists began thinking about a fusion bomb.

In an atomic bomb, fission fuel such as plutonium or an isotope of uranium is brought together and compressed by the force of a conventional chemical explosion. Under these conditions the plutonium reaches what is called a "critical mass," and an uncontrolled chain reaction results. In an incredibly short time, nuclei begin to split and at the same time release their energy. The result is the explosive release of heat, light, and nuclear radiation of an atomic bomb. But, scientists wondered, why not go one step further and work with nuclear fusion instead of fission?

In fact, in the early months when the first atomic bombs were being designed at Los Alamos, there was some debate that their explosion would actually cause a nuclear fusion. Until the first A-bomb had been exploded, no one had produced such incredible temperatures on earth. Some scientists began to worry that these temperatures would be high enough to trigger nuclear fusions in water. What if all the hydrogen in the world's oceans were to begin to fuse? The exploding of a single atom bomb would be sufficient to burn up the whole earth.

Eventually, careful calculations assured the Los Alamos physicists that uncontrolled fusion would not be possible. (For a time, however, the future of the globe hung on the reliability of a scientific theory.) But soon a fusion bomb had become a distinct possibility, and Edward Teller in the United States and Andrei Sakharov in the Soviet Union began working out its fine details. The idea would be to ignite a mass of hydrogen and cause its nuclei to fuse and release mass-energy.

The fusion reaction in a hydrogen bomb is triggered by the explosion of a conventional atomic bomb held inside the H-bomb casing. The A-bomb itself relies for its energy on a runaway chain reaction involving nuclear fission of plutonium, and the shock wave of energy from the exploding A-bomb is focused onto the fusion fuel inside the H-bomb casing. Under intense pressures from the A-bomb and a temperature around 25 million degrees, the nuclei in the fuel fuse, releasing some of their mass as energy. The whole process takes place in an incredibly short time and leads to the explosive release of heat, light, and radiation. Through the annihilation of a tiny amount of mass, science has been able to release the potential for destroying life on this planet.

CONTROLLED FUSION

Nuclear fusion, whether in the heart of a star or in a hydrogen bomb, involves temperatures of tens of millions of degrees. If fusion is to be harnessed on earth—or at least so physicists believed until Jones, Fleischmann, and Pons came along—then laboratories must duplicate the conditions within the sun or the H-bomb.

This problem of producing, containing, and maintaining a fusion reaction at tens of millions of degrees has preoccupied physics for many decades. It is this same problem that Jones, Fleischmann, and Pons were attempting to bypass in an ingenious way.

On March 23, Fleischmann and Pons announced that controlled fusion had been carried out in a test tube. The approach they had used was truly revolutionary. To understand just what they claimed to have done, it is first necessary to learn something about the more conventional approach to regular nuclear fusion and the experiments that are currently being carried out in high-temperature plasma fusion laboratories throughout the world.

HYDROGEN ISOTOPES

Nuclear fusion involves the marriage of two hydrogen nuclei along with a corresponding release of energy. The fuel used by our sun, and by high-temperature fusion laboratories on earth, is hydrogen. It makes good sense to continue this exploration of nuclear fusion by discovering just what goes on in an atom of hydrogen when it fuses.

The hydrogen atom consists of a tiny central positive core, or nucleus, with a single negative electron outside. It is this electron that

Figure 2-3

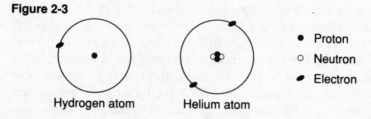

Hydrogen atom   Helium atom

● Proton
○ Neutron
🖋 Electron

gives hydrogen its particular chemical properties—for example, its high reactivity with oxygen. Thanks to this single outer electron, hydrogen is able to react with oxygen and form the most abundant substance on earth—water, or H_2O.

The atomic neighbor to hydrogen is helium, with two electrons outside a positive central core. The pairing of two electrons in helium makes this gas extraordinarily stable and unreactive. Unlike hydrogen, helium does not engage in chemical reactions and does not generally form compounds. Helium cannot combine with oxygen, for example, to form a chemical analogue of water.

Hydrogen seems to envy the stability of its helium cousin, to the extent that its single electron is constantly trying to find a friend to pair up with. In fact, when two hydrogen atoms meet, their electrons pair up to create a hydrogen molecule, H_2. This is the normal form in which hydrogen gas exists—as a two-atom molecule, rather than as single isolated atoms.

Figure 2-4

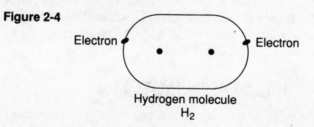

Electron   Electron

Hydrogen molecule
H_2

This activity of hydrogen's electrons will become important in the next chapter's discussion of cold fusion. In fact, the way electrons behave around a nucleus of hydrogen is the whole key to the Utah process. But, for the present, let us focus on the nucleus of the hydrogen atom, not on its electron.

Provided that the hydrogen nucleus bears a single positive

charge to balance the charge of the outer electron, chemically speaking it does not much matter what else it contains. As far as chemists are concerned, it will continue to behave like hydrogen.

However, the nuclei of hydrogen atoms are not all identical. Although they all bear a single positive charge, their masses can in fact differ. Of all the hydrogen in the universe, the vast majority of it exists in the form of what is called "light hydrogen." In this form, the nucleus consists of a single elementary particle called a proton. The proton has a single positive charge. (The electron has an exactly equal and opposite negative charge, but its mass is some 2,000 times lighter than that of the proton. Nearly all the mass of a normal hydrogen atom therefore resides in its nucleus.)

Figure 2-5

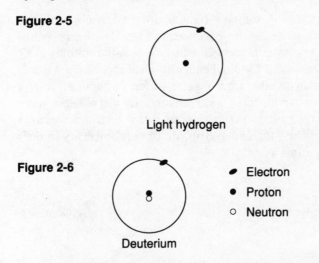

Light hydrogen

Figure 2-6

Deuterium

Electron
Proton
Neutron

Another form of hydrogen is called heavy hydrogen, or deuterium. One in every 6,700 atoms of hydrogen exists in this heavy form. It has a nucleus that is roughly twice as heavy as that of normal hydrogen. While the nucleus of ordinary light hydrogen contains a single positively charged proton, that of deuterium contains a proton plus a neutron. The neutron is approximately the same weight as the proton but has no charge. Chemically speaking, it is simply ballast and has virtually no effect on the chemical properties of heavy hydrogen.

Deuterium is called an *isotope* of hydrogen and is chemically identical to ordinary light hydrogen. That is, deuterium enters into exactly the same chemical reactions and combinations as ordinary light hydrogen. The only difference is that its reactions tend to be a little slower.

Figure 2-7

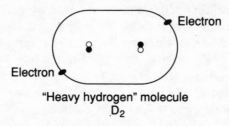

Electron

Electron

"Heavy hydrogen" molecule
D_2

Just as atoms of ordinary light hydrogen combine to form molecules of H_2, so deuterium combines to form D_2—heavy hydrogen gas. Ordinary water is represented by the familiar formula H_2O; heavy water consists of D_2O. Chemically speaking, heavy water is virtually indistinguishable from regular water. (Although bacteria have been grown in pure heavy water, an exclusive diet of heavy water would be lethal to humans.) It is this same heavy water that was used at the University of Utah and at Brigham Young University in their cold-fusion experiments.

Figure 2–8

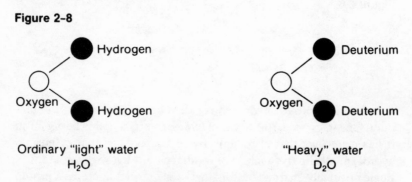

Hydrogen

Oxygen

Hydrogen

Ordinary "light" water
H_2O

Deuterium

Oxygen

Deuterium

"Heavy" water
D_2O

By a painstaking physical process, the tiny percentage of heavy water can be extracted from ordinary tap water. A very slight difference in their densities and boiling points distinguishes the two forms of water. The precision required to measure these differences makes it a tricky matter to separate heavy water from light water. Consequently, bottle for bottle, heavy water is more expensive than the best Scotch whiskey. Although heavy water is abundant in nature, a cost must still be paid for extracting it from lakes and oceans.

Figure 2-9   COMPARING WATER

Light (or ordinary) water is composed of molecules containing two hydrogen atoms and one oxygen atom, in chemical notation written H_2O.
A molecule of heavy water contains deuterium—the heavy isotope of hydrogen—and oxygen and is written D_2O.

	Ordinary water	Heavy water
Freezing point (° C)	0.00	3.81
Boiling point (° C)	100.00	101.42
Density at 25° C (g/cm³)	0.99701	1.1044
Viscosity at 55° C (mPa.s)	0.8903	1.107
Molecular weight	18.015	20.028

Heavy water is used in scientific experiments and in the nuclear power station (CANDU) manufactured by the Canadians. For this reason, Canada has become one of the world's major producers and exporters of heavy water. It may be no coincidence that Ontario Hydro, which produces heavy water and runs the CANDU reactors, was the first corporation to sign an agreement with the University of Utah.

To complicate matters still further, and to fill in the background needed for this fusion story, there is yet one more heavy isotope of hydrogen, called tritium. Tritium is particularly rare. In fact, only one in every 1,000,000,000,000,000,000 hydrogen atoms exists in the tritium form.

The tritium nucleus is three times heavier than that of ordinary light hydrogen. It contains one proton and two neutrons. The second neutron increases the mass of the tritium nucleus, but it does not make for a particularly stable arrangement. In fact, tritium is radioactive; it gives off beta radiation. Tritium is said to have a *half-life* of twelve and a half years, which means that there is a fifty-fifty chance that a tritium nucleus will decay in that time.

Figure 2-10

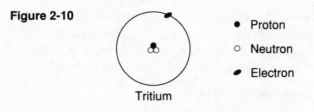

- • Proton
- ○ Neutron
- ✎ Electron

Tritium

It turns out that, in the conditions generated within a thermonuclear bomb, the tritium nucleus is a lot easier to fuse than a nucleus of light hydrogen. For this reason, tritium rather than light hydrogen forms the fuel of a hydrogen bomb. Because of constant radioactive decay, H-bombs do not last indefinitely but must be periodically "refueled."

Figure 2-11

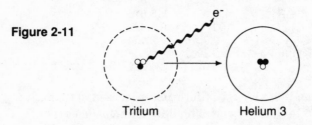

Tritium   Helium 3

Tritium decays by emitting a ß-ray (an electron) and transmuting into Helium 3. The "half-life" of this reaction is 12.26 years.

Just as deuterium combines with oxygen to form water—"heavy water," or D_2O—so tritium combines to form what could be called super heavy water, or T_2O. An incredibly tiny amount of the world's water is in this form, certainly too little to extract by normal means. But tritium can be recovered in other ways.

It forms naturally within the heavy water of a CANDU reactor. Over many months, the heavy water (D_2O) inside the nuclear fission reactor is bombarded by the neutrons that are produced in the fission reaction itself. These neutrons enter the deuterium nucleus (one proton and one neutron) and convert some of the deuterium into tritium (one proton and two neutrons). Tritium can therefore be periodically extracted from the heavy water of a CANDU reactor. The Fusion Fuels Technology Project in Ontario, Canada, has become the world's major producer and handler of tritium.

Tritium is also produced commercially by bombarding lithium with neutrons. In the United States, the tritium for H-bombs was generated by placing lithium within a fission reactor so that the neutrons could transmute it. In addition to its core of tritium, a hydrogen bomb contains a cladding of lithium. Nuclear fusion of the tritium releases neutrons that act on the lithium to produce yet more tritium. In its act of explosion, a hydrogen bomb is actually generating more of its own fuel.

Fusion Reactions

With three hydrogen isotopes in nature, physicists have several reactions to choose from when they plan to harness nuclear fusion. It turns out that mimicking the reactions inside the sun by fusing the nuclei of light hydrogen is, as a practical matter, far too difficult. Fusion physicists have to content themselves by working with the heavier isotopes of hydrogen, and they must still be willing to generate enormous temperatures in the laboratory.

The reason it is so difficult to produce fusion in the laboratory, or for that matter in the sun itself, is that the two nuclei must first be brought close enough together to react. In addition, they must spend enough time together to allow this reaction to lead to actual fusion. It seems that nature often hides its secrets and puts up a barrier to our endeavors. It is just this barrier to easy fusion that has forced scientists and engineers to create major research laboratories and giant machines capable of bringing hydrogen nuclei close enough together to react.

The problem with getting nuclei close enough together is the positive charge on the nucleus. Like charges repel each other. Getting two nuclei side by side is like pushing together the north poles of two magnets. The closer the poles are forced together, the greater the repulsion between them. Until Jones, Fleischmann, and Pons came along, physicists believed that the best way to get two hydrogen nuclei close enough to fuse was to smash them into each other at very high speeds. When two nuclei speed toward each other, there is a chance that, for an incredibly short fraction of a second, they will get close enough to react.

Figure 2–12

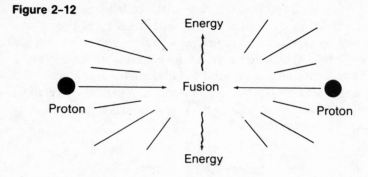

The hotter a gas, the faster its atoms are moving. The clue to conventional nuclear fusion therefore lies in getting the hydrogen fuel so hot that its nuclei are going fast enough to collide and fuse. In addition, a second condition must be met: the two nuclei must spend long enough together for them to react. Otherwise, like two billiard balls, they will bounce away as if nothing had happened.

What sorts of temperatures are needed for a nuclear fusion? Tens of millions of degrees, temperatures similar to those at the center of the sun. This is no coincidence, since physicists are attempting to duplicate this very reaction in the laboratory. It turns out that the temperature required to produce a fusion reaction depends upon the isotope used. The following table shows that the lowest temperatures for fusion occur with a deuterium and a tritium nucleus.

We shall come back to the reaction between two deuterium nuclei later, for this is exactly the reaction that first aroused the interest of the cold-fusion scientists. But in this chapter we are focusing on high-temperature fusion, and this means reacting deuterium and tritium together at a temperature of at least 45 million degrees. To attempt anything else requires pushing up the temperatures by a factor of ten.

The preceding figure suggests that if a mixture of deuterium and hydrogen is heated to 45 million degrees, the nuclei involved will be moving so fast that when they collide, a fusion reaction can take place.

The magnitude of the problem facing the fusion community is quite clear. How can such extremely high temperatures be generated

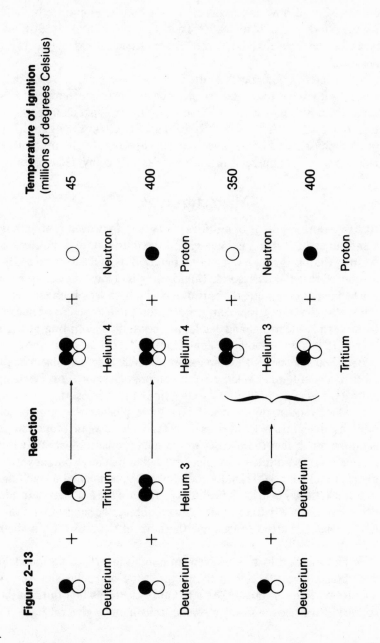

Figure 2-13

on earth? And how can a gas at such a high temperature be held together long enough to react? These are the incredibly difficult problems that have taxed the ingenuity of physicists and engineers for decades.

To date, various parts of the problem have been solved. Some laboratories have managed to produce the extreme temperatures involved; others have been able to contain the hot gases long enough. But to get all the factors exactly in place requires some extremely precise physics and some ingenious engineering. In the opinion of many experts, commercial fusion power is still many decades away.

Magnetic Bottles

At temperatures of tens of millions of degrees, hydrogen is not really a gas but rather what has been called the fourth state of matter—a plasma. Until this century, scientists believed that all substances exist in one of three states: solid, liquid, or gas. Now physicists have learned of a fourth state, the plasma, in which, at very high temperatures, electrons are stripped away from atoms to leave charged nuclei. In a plasma, electrons and nuclei are constantly colliding at high speed. It is this plasma of deuterium and tritium that must be maintained at a very high temperature and held together long enough for fusion energy to be released on a commercial scale. This is where an ingenious concept, called a magnetic bottle, comes in.

Clearly no material could contain a plasma that is tens of millions of degrees hot. No glass or metal could serve that function. It is not so much that the plasma would melt the metal—actually the plasma is far too dilute to do this—but rather that any contact with a metal container would rapidly cool down the plasma and quench the fusion reaction. The trick therefore is to contain the hot plasma inside the reaction vessel without the plasma touching the sides in any way. In a sense, scientists must hold the core of the sun within their laboratory.

The solution to this problem of containment can be found in your home. The clue lies in the picture tube of a television set. A television picture is created by a beam of electrons that falls on the surface of the phosphorescent screen, causing it to give off light. To create a recognizable picture, this beam must be made to move around in obedience to the signal that arrives from the transmitter. The trick in doing this is to create a magnetic field that bends and moves the beam. Electrons, being electrically charged, can be made to change their paths in a magnetic field. Varying this magnetic field permits the beam of electrons to be moved at will.

The electrons in the picture tube are a little like the charged nuclei in a high-temperature plasma. This suggests that it may be possible to arrange a very strong magnetic field in such a way that the plasma can be moved around at will. In fact, the goal is to trap the plasma within the field; in other words, the magnetic field can be made to act like a bottle.

There are two ways in which a hot plasma can be trapped by a magnetic field. In one of these, the plasma has a sausage shape. Charged nuclei shoot along the sausage, where they are reflected by a "magnetic mirror" at the end. In the other approach, the nuclei are confined in a doughnut shape by powerful magnets. This latter kind of magnetic confinement, called a *tokamak*, is currently the most popular design in fusion laboratories. One of the architects of the tokamak is Andrei Sakharov, father of the Russian H-bomb and winner of the Nobel Peace Prize for his work on civil rights.

In a tokamak reactor, powerful superconducting magnets confine the deuterium and tritium nuclei within the doughnut-shaped reactor. All that remains is to pump the plasma to a high enough temperature and continue to confine it while the fusion reaction takes place.

All this is easier said than done, for achieving the break-even point in nuclear fusion is a combination of physics principles and engineering skills.

Two magnetic fields must be used to contain the hot plasma. One of these confines the plasma to move around the axis of the doughnut. While this holds the hot plasma together, it still allows the plasma to wander and drift sideways so that it can hit the walls of the tokamak. A second set of magnet windings is therefore added to produce a circular field that holds the plasma in the center. The combined result of these two fields is to make the nuclei spiral along the doughnut shape.

In principle, the tokamak and the magnetic mirror solve the problem of how to hold the plasma together at a temperature of tens of millions of degrees. In practice, the process may be more complicated, but although many technical problems remain to be solved, it is at least the first step toward harnessing high-temperature nuclear fusion.

Heating the plasma to tens of millions of degrees is achieved in a variety of ways. For example, a beam of particles traveling at very high speeds can be fired into the plasma. This causes collisions, which heat up the fuel. Another approach is to use microwave heating. As with a piece of beef in a microwave oven, the temperature of the plasma can be pumped higher and higher. The only difference is the amount of power involved.

The giant magnetic field that is used to confine the plasma within a tokamak also plays a secondary role. Varying this magnetic field causes an electrical current to flow. This is the principle on which every electrical generator works. In the case of a tokamak, the varying magnetic fields generate electrical currents that circulate within the plasma itself. The effect is exactly like what happens in the electrical heater in your home, only this time it is not a wire that heats, but the plasma itself.

A combination of heating by firing in beams of neutral particles, microwaves, and currents in the plasma itself can raise the temperature inside the plasma to tens of millions of degrees. A record temperature of 200 million degrees has been achieved at the Princeton Plasma Physics Laboratory. But even this was not sufficient to produce fusion power. The key is to create the combination of exactly correct conditions in which high temperatures are maintained and the plasma itself is contained long enough.

In addition to these two important factors, enough plasma must be present. In a tokamak, plasma densities are incredibly low, only one hundred-thousandth of the density of air here on earth. The problem with such low densities is that far fewer nuclei collide each second. Since not every collision necessarily leads to fusion, the process is not very efficient. If the pressure could be increased, then more nuclei would collide each second, more fusing reactions would take place, and more energy would be released.

The key to achieving usable nuclear fusion therefore lies in a combination of three factors: getting high enough temperatures, holding the plasma together for a long enough time, and achieving high enough plasma densities. Some of the large experimental machines that are at present located in Europe, the United States, the Soviet Union, and Japan can achieve one of these conditions, but none can yet produce them all at once.

This critical combination of factors is called the Lawson condition, and that condition is the goal that fusion laboratories are striving for. It is named after John D. Lawson, a physicist at Harwell, England, who first pointed out this combination of conditions in 1957.

Fusion Milestones

Turning high-temperature nuclear fusion into a worldwide supply of power will not be achieved overnight. In fact, a series of milestones must be passed on the way toward commercial high-temperature fusion:

Scientific breakthrough	The fusion energy released is equal to the power needed to maintain the hot plasma.
Ignition	Enough energy is released to sustain the plasma reaction without a need for further heating. Once started, the fusion reaction maintains itself.
Engineering breakeven	The total fusion power is greater than the power needed to run the plant.
Commercial viability	Fusion power competes favorably with other forms of energy.

Some research groups believe that the scientific breakthrough in fusion power is just around the corner. Even so, fusion's commercial competition with nuclear fission may be many decades away.

Inertial Confinement

An alternative approach to the magnetic bottle that is in use involves what physicists call *inertial confinement*. Inertial confinement attempts to duplicate the conditions inside a hydrogen bomb by simultaneously firing intense laser beams of energy from many different directions onto a pellet of fuel.

Powerful lasers are arranged around a spherical chamber into which a tiny frozen pellet of fuel is dropped. The instant this falling pellet reaches the center of the chamber, a spherical bank of lasers fires in unison. In less than a billionth of a second, the radiation from these lasers compresses the surface of the pellet to a pressure 100 million times that of atmospheric pressure. The outer layers of the fuel reach a temperature of up to 100 million degrees and a density twenty times that of lead.

For around one ten-billionth of a second, the conditions deep in the heart of a star are reproduced, and fusion energy is released. Then the reaction is over, a moment later the second pellet falls, the lasers fire again, and the whole process is repeated.

THE FUTURE OF FUSION

The bottom line on fusion power is a matter of economics. Fusion laboratories—whether they be tokamaks, magnetic mirrors, or lasers firing at frozen pellets—are megaprojects and require megadollars. In fact, over the last decade, it has become apparent that harnessing nuclear fusion is so complex that no one nation on earth could do it alone. In a historic meeting, the leaders of the United States and the Soviet Union agreed to cooperate in fusion research at the international level. To avoid tipping the balance of power, they also agreed that the first fusion reactor would not be built in either country but in some third territory.

The next phase was a joint agreement among the United States, the Soviet Union, Japan, and the European Community to work toward an International Thermonuclear Experimental Reactor

(ITER). Later Canada joined in this venture under the auspices of the European Community. At present all these countries have fusion programs of their own but have agreed to pool their data and work on common experiments. The long-term goal is to design and build a large fusion reactor that will work under conditions that simulate those expected in a commercial plant. The ITER is planned to have a diameter of 10 meters and generate 1,000 megawatts of fusion power (the power of the biggest fission reactors) with electrical currents of over 20 million amps circling the plasma.

But fusion research has not gone as rapidly as everyone had hoped, mainly because of recent budget cuts in the United States. Funding for magnetic-bottle experiments, for example, dropped from $362 million in 1986 to $331 million in 1987. Work on inertial confinement continues at the Lawrence Livermore, the University of Rochester, and, with a modified approach, the Sandia national laboratories. But it will be a number of years before fusion is ignited.

Commercial Fusion

With high-temperature fusion successfully harnessed sometime in the future, how will a commercial fusion reactor operate? Essentially, the intense heat produced in the fusion reaction will be converted to electricity. Although temperatures of tens of millions of degrees are involved, the plasma in present tokamaks represents only a tiny wisp of gas. Its density is closer to that experienced at the edges of space than that on even the highest mountain. The key to greater tokamak power will be to produce a plasma of a high enough density to yield a usable amount of heat.

In the first machines, a closed-loop cooling system will remove this heat from the tokamak and pass it on to a series of heat exchangers. In turn these heat exchangers will be used to produce steam that drives electrical turbine generators. In this respect, apart from the unique nature of the reactions involved, a fusion power plant will be similar to a nuclear fission reactor.

In the next generation of fusion systems, the cooling system will consist of helium gas that will drive gas turbine generators. The result will be a far more efficient way of producing electricity from heat. Scientists are even talking about advanced systems that will produce electricity directly from fusion power without the intermediate steps of heat exchangers and electrical generators.

But the energy of a tokamak need not be used only to create electrical energy. Because they are safer than present nuclear fission reactors, fusion reactors can be built closer to large cities. This opens the possibility of using fusion plants to heat homes and offices. Close to a town, the tokamak's heat will be conveyed by steam pipes. For slightly longer distances liquid salt can be used to pump the heat, while for distances up to 100 miles a chemical heat pipe will be needed. The heat from a commercial tokamak could also be used to desalinate water and drive various processes in a chemical plant.

Unlike conventional nuclear power stations, tokamaks do not suffer from the problem of radioactive spent fuel. The used fuel from a conventional nuclear reactor is highly radioactive and, for decades and even centuries, must be prevented from escaping into the environment. Tokamaks do not produce such radioactive waste even though tritium is used as a component of the fuel. Tritium, although a radioactive hazard, decays in a matter of years and is relatively easy to handle.

There is, however, nuclear radiation associated with the fusion process itself, but experts are already thinking of turning this to an advantage. With each fusion reaction, a single neutron is emitted. This neutron is an elementary particle that carries no charge and has the effect of making other materials radioactive. The neutrons emitted from the fusion reaction will cause the construction materials within the tokamak to become radioactive.

However, by using special construction materials, like ceramics, it may be possible to minimize this radioactivity and confine it to isotopes that are short-lived. The neutrons that are emitted from a tokamak can also be used to create radioisotopes for commercial use, such as in industry and medicine. For example, by cladding the tokamak in cobalt metal, it will be possible to create cobalt 60, an isotope that is in demand all over the world for cancer treatment. Cobalt 60 is used in the so-called cobalt bomb used in radiation treatments. Cobalt 60 and other radioisotopes can be used for

Figure 2–14   COBALT 60 DEMAND

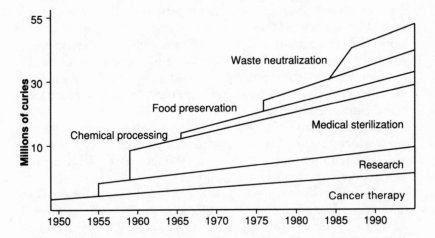

The demand for cobalt 60 will increase toward the end of the century as new uses, such as food sterilization and the neutralization of toxic waste, become important.

sterilizing food and for taking x-ray photographs of metals such as on aircraft to look for fractures.

In fact, a whole range of industrial radioisotopes will be produced by the fusion power stations of the future. Rare metals such as rhodium, selling at $1,200 per ounce and used for electrical contacts and chemicals catalysts, and osmium, used to make hard alloys and in biochemistry, will be created through neutron bombardment of cheaper metals.

A particularly important product to emerge from a high-temperature fusion plant will be hydrogen. Hydrogen is a component of ordinary water (H_2O). It is made by electrolysis (passing an electrical) current through water) or by a chemical reaction. In both cases, high temperatures help split water into its components of hydrogen and oxygen. With the plentiful heat and energy available from a tokamak, it will be relatively easy to produce large volumes of hydrogen. In fact, many futurists are already talking about a "fusion-hydrogen" economy, in which part of the energy from a tokamak is used to make hydrogen—the fuel of the future.

A tokamak produces heat and electricity, both of which are needed in homes and factories. But what about driving your car to work? Will this activity still rely upon rapidly depleting reserves of gasoline? One scenario for the automobile of the future is to use a transportable energy fuel as a replacement for gasoline. Hydrogen fits the bill; in fact, it does a far better job. Pound for pound, hydrogen contains more energy than gasoline.

It is also a far cleaner fuel. The only product of burning hydrogen is water—steam. The process does not produce carbon dioxide and therefore does not add to the greenhouse effect. It does not produce other contaminants or acid rain.

Hydrogen is a concentrated energy form and can also be used as the starting point in the synthesis of alternative fuels like methane, methanol (wood alcohol), and synthetic gasoline. Experts project a fusion future in which the burning of fossil fuels is eliminated. Heat and electrical power will be produced at fusion plants, with hydrogen and other synthetic fuels used in transportation and as the starting point in the synthesis of new chemicals.

The Economics of Energy

The world has many forms of energy production, from wood stoves to nuclear fission power reactors. In the future, high-temperature fusion power will have to compete economically with these energy sources. Several decades into the next century, if high-temperature fusion power becomes a reality, its economics will have to be assessed against those of rapidly dwindling fossil fuels. Possibly by that time fission will be considered safe, and the problems of waste disposal will be solved. On the other hand, pressures by environmental groups and other concerned citizens may have closed down the last nuclear plant.

Current estimates suggest that there will be a 2 percent growth in energy demand—which means sixty new large power plants, either conventional or fission, in the United States by 2040. Since a large *fusion* reactor could produce as much power as twenty nuclear power stations, fusion appears to be an important part of the U.S. energy future.

In assessing the economics of each of these energy supplies, a number of costs have to be taken into account:

- The cost of building the power plant

- The cost of the fuel

- The cost of operating and maintaining the plant

- The cost of the social and environmental impact of the operating plant, including its various safety features and the possibility of accidents or cleanups

- The cost of decommissioning the plant

While fossil fuels are expensive and unacceptably stress the environment, a coal- or an oil-fired power plant is cheap to build and easy to maintain. On the other hand, nuclear fission power begins with cheaper fuel, but the reactors involved are more expensive to build. Fusion plants will be even more costly to construct than fission reactors but will use the cheapest of all fuels.

All power plants present a hazard when they work at high temperatures and contain so much energy in a small region. Take, for example, a nuclear power plant. It is relatively simple to switch off the chain reaction in a nuclear power plant. Even in the event of a very serious accident, the reactor can be shut down. The problem is that, even with the chain reaction successfully quenched, so much energy is stored in the fuel rods, in terms of their radioactivity, that they quickly begin to heat up. If there is a major failure in the cooling system, then these fuel rods will heat so much that their casing will begin to distort, crack, and even melt. The result can be a release of radioactive fuel into the environment or even what is called a meltdown—a critical condition in which white-hot molten fuel burns its way through the reactor floor and into the bedrock beneath.

The nuclear accidents at Three Mile Island and at Chernobyl were essentially of this nature. It was not so much a runaway nuclear reactor that caused radioactive contamination to be released; indeed, the operators were able to shut down the reactions with ease. Rather, the fuel itself began to generate heat through normal radioactive decay, which led to a partial meltdown.

This sort of accident could never happen with a tokamak. To begin with, the fusion reaction can take place only at a temperature of tens of millions of degrees, and this must be actively maintained. Any breakdown in the system would lead to immediate cooling. For example, if the magnetic containment system were to fail, then the plasma would drift, touch the walls of the reactor, and immediately cool.

But what about the effect of such high temperatures on the walls of the tokamak? Could the walls not explode and destroy the plant? The answer is that plasma itself is so tenuous, and at a density hundreds of thousands of times less than the air we breathe, that the temperatures would have a negligible effect.

If an explosion or other large release of fusion energy is impossible, then what other safety hazards would a commercial tokamak present? There is the matter of radioactivity. The fuel itself is relatively safe. Tritium does happen to be radioactive and is a serious health hazard. But because its half-life is only about twelve and a half years, it does not present the sort of long-term hazard that is associated with the spent fuel from nuclear fission. As a gas, tritium would quickly dissipate into the atmosphere, where it would become diluted. In addition, the fusion plasma is so diluted that very little tritium is present. Most of the tritium would be in storage at the power plant. The major hazard from tritium is that the gas can easily combine to form a radioactive cousin to heavy water. If it were to contaminate a water supply in this way it could be easily absorbed into the human body.

Clearly the research and development involved in creating high-temperature fusion are extremely costly, and it will take many decades before a commercial system is built. The fusion plants themselves will represent a considerable capital investment. These reactors will be megaprojects, in which large amounts of energy are produced in a single locality and then transported to the surrounding area. For those who believe that the world should be moving toward what is called appropriate-sized technology, megaprojects like fusion do not look so attractive.

The first fusion reactors will probably be so costly they will be built by the richest nations. Will that further the gap between the rich and poor? High-temperature fusion power may well solve the energy problem—but initially that solution may be offered only to the very rich. A few huge fusion reactors dotted across the world may also mean a new kind of centralized power, and this may create political problems.

Cold-temperature fusion is exciting because it offers a radically different possibility. If small cold-fusion reactors could be made to work and if the energy they produce is cheap enough, then cold fusion would revolutionize the whole politics of energy. Imagine a cheap energy source; imagine fusion reactors that are not capital-intensive and a technology that is available at a scaled-down level. There is even talk about cold-fusion cells in every home and in automobiles. Cold fusion would be available to everyone, to all nations.

In the final chapter, we will return to some of these social and economic considerations, including speculation about the energy future our children will have to contend with.

Chapter 3
Muon-Catalyzed Fusion

"We had a short but exhilarating experience when we thought that we had solved all the fuel problems of mankind for the rest of time. While everyone else had been trying to solve this problem by heating hydrogen plasmas to millions of degrees, we had apparently stumbled on the solution, involving very low temperatures instead."

No, this statement did not come from Pons and Fleischmann at their press conference on March 23. The events referred to happened over thirty years earlier. The speaker is a physicist, Luis W. Alvarez. Alvarez later became more widely known for his idea that the sudden extinction of the dinosaurs was caused by a giant meteor that hit the earth. But at the time of this statement, Alvarez was running experiments at the Los Alamos Meson Physics Facility. And for a short, exhilarating period, he believed that he had discovered the clue to cheap and abundant energy. The process he had stumbled upon was later to become known as cold fusion.

For several decades, physicists had assumed that nuclear fusion could take place only at extremely high temperatures inside apparatus that duplicates conditions inside the sun. But in analyzing a quite separate experiment, Alvarez had noticed a puzzling new phenomenon. In the end, it could only be explained by assuming that room-temperature fusion was taking place. The result was a great shock, and for a time Alvarez and his group believed that they had solved all the world's energy problems. It was only when they got down to making some hard calculations that they realized that the amount of cold fusion occurring could not compete as an energy source with the more familiar forms like coal, oil, gas, fission, and hydroelectric power.

Cold fusion, as Alvarez had seen it, would never be economically important. It was only decades later that cold fusion was to hit the headlines. At the time, the process first observed by Alvarez remained something of a scientific curiosity. It was not until May 23, 1989, that the idea of cold fusion again burst upon the world as scientists from the University of Utah and Brigham Young University claimed they had worked out a totally new way of achieving something that had been first hinted at thirty years before.

Fusion happens when two nuclei of hydrogen meet. As we have seen, there are a number of possible fusion reactions involving various combinations of the isotopes of hydrogen (light hydrogen, deuterium, and tritium). The problem is that nuclei repel one another, and it takes a great deal of effort to bring them close enough to react. The solution that nature has adopted in the sun and stars is to work at temperatures of tens of millions of degrees. At such high temperatures, the nuclei are moving very rapidly, so fast that when they hit each other, they can approach closely before their mutual repulsion takes over and forces them apart again.

But even at a temperature of tens of millions of degrees, two hydrogen nuclei are still not moving quickly enough to guarantee a reaction every time they hit. Theory shows that their point of closest approach is still too distant to allow fusion to take place.

How, then, is it possible for the sun to shine? How can fusion ever take place if the individual hydrogen nuclei never come close enough together? The answer lies in a curious phenomenon that was first described by Werner Heisenberg in the same year that he created the new science of quantum theory. Heisenberg discovered what has become known as the uncertainty principle, which says that, at the elementary particle scale of things, there exists a basic uncertainty or ambiguity in nature.

Heisenberg showed that, when it comes to electrons or nuclei, there is an uncertainty in some of their properties. For example, if you are able to pin down the speed of a nucleus, there will be some degree of uncertainty in your determination of its position. If you can pin down its position, then you will be uncertain of its speed.

Figure 3-1

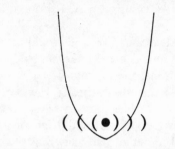

An elementary particle at the bottom of a well has some uncertainty in its position.

This uncertainty has nothing to do with human ignorance or with our inability to carry out accurate measurements. It is a fundamental property of the universe that tells us there is an intrinsic ambiguity in pinning down the specifications of the nucleus.

When two hydrogen nuclei collide at 10 million degrees, logic tells us that they still are too far apart to react. But now the uncertainty principle suggests that there is actually a basic uncertainty at this point of closest approach. Possibly they are a little closer than we had guessed, possibly they are a little farther apart. All these various possibilities must be taken into account, including one that suggests that they may actually be close enough to fuse and release energy.

Thus, quantum uncertainty tells us that when two speeding nuclei collide, there is an element of uncertainty in their point of closest approach and hence a small probability that they are actually close enough to fuse. Nature has provided a curious trick whereby the impossibility of nuclear fusion is transformed into a slim probability. If it were not for this trick, the sun would not shine and the stars would not twinkle in the night sky.

Physicists generally describe this sleight of hand on nature's part as "quantum tunneling." Think of the repulsion between two nuclei as a barrier they must mutually overcome. The following diagrams show what is going on. The faster the nuclei travel, the greater their chance to go through the barriers. But even at 10 million degrees, they do not have enough speed to get to the top of the barrier and react.

Figure 3-2

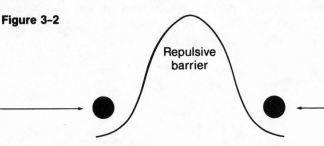

Two speeding nuclei do not have sufficient energy to overcome the repulsion between them.

Quantum tunneling suggests another way of bypassing the barrier—not climbing over it, but tunneling through it. Quantum uncertainty suggests that there is always a finite but small probability that the nucleus will find itself in another location, at the other side of the barrier that separates them. Another way of talking is to say that the two nuclei have a small but nonzero probability of tunneling through the barrier. This, in fact, is the way that nuclei in a high-temperature plasma interact. If they can at least get close enough together and for a long enough time, they are able to complete the process by tunneling through the repulsive forces that hold them apart.

Figure 3-3

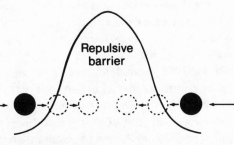

There is a slight probability that two speeding protons will tunnel through their repulsive barrier and fuse.

Quantum tunneling is like a coin toss—there is always a small chance that the two nuclei will interact before bouncing apart again. As the temperature increases, the nuclei collide faster, they get closer, and they have a greater probability of tunneling through to each other and reacting. In other words, one way of improving a nucleus's chance of reacting is to raise the temperature.

Figure 3-4

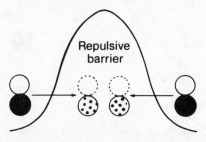

Two deuterium nuclei are held close to the repulsive barrier. Although the chance of quantum tunneling is very low, by waiting long enough the two nuclei may eventually meet and fuse.

But there is another and more subtle strategy of achieving fusion. Suppose that the two nuclei approach, not as close as before, but this time do not bounce apart again. Suppose that they are held together, at a greater distance apart but for a much longer time. There will always be a very small probability that the nuclei can tunnel through to each other. This probability is very small indeed, but the longer the nuclei are held in close proximity to each other, the greater is the chance that they will react and fuse. This second, gentle route was the one pursued by the two groups in Utah.

The problem was, of course, to discover some ingenious way of bringing two nuclei close together and then holding them long enough for them to tunnel. At room temperature in an ordinary gas of hydrogen or deuterium, the nuclei are so far apart that even if one waited for billions of years, quantum tunneling would still not take place. Somehow the hydrogen or deuterium nuclei had to be persuaded to approach even closer and to remain there while quantum tunneling took place.

In summary, therefore, there are two ways in which fusion could be brought about:

1. High-temperature fusion—This involves heating a plasma to tens of millions of degrees so that the nuclei approach close together for an incredibly short fraction of a second.

2. A mysterious second way in which the nuclei do not approach as closely but are held together much longer—This is the way of cold fusion.

Figure 3-5

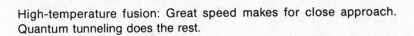

High-temperature fusion: Great speed makes for close approach. Quantum tunneling does the rest.

Room-temperature deuterium: The nuclei are too far apart for quantum tunneling to have any effect.

Cold fusion: Discover a way to bring the nuclei closer and hold them there.

The first way is currently being attempted in tokamaks around the world. The second is far more speculative and, until Fleischmann and Pons at Utah and Jones and his group at Brigham Young University came along, it had never been a serious contender for fusion energy.

At this stage of the argument, fusing two nuclei of hydrogen looks like a curious theoretical possibility, but could it ever be achieved in practice? How could nuclei be persuaded to come together? The answer, when it came, was a pure accident.

In 1956, Luis W. Alvarez and his colleagues did not have cold fusion in mind when they began to look at the results of the experiments they had been carrying out at the Los Alamos Meson Physics Facility. These experiments involved a new particle accelerator, which accelerates elementary particles to very high speeds so that their properties can be studied when they collide with other particles. This had nothing at all to do with practical nuclear fusion but was simply a way of discovering the forces that act among the elementary particles and the various ways they behave.

In some of the experiments, a beam of particles called muons was created. For experimental reasons, this beam passed into a container of liquid hydrogen and liquid deuterium. It was purely by chance that anything out of the ordinary was discovered. The scientists were actually looking at the tracks that the particles left on pieces of film. As they began to look more closely, they noticed some puzzling and unexpected patterns.

What they observed was so unusual as to baffle Alvarez and his colleagues. There was a clear indication of a muon coming in—as everyone had expected. Then a gap. Then a new muon track that lasted for exactly 1.7 cm. In photograph after photograph, the physicists clearly saw the path of the incoming muon ending in a gap and then followed by a path that was 1.7 cm long. What on earth could it mean? What was the special significance of that distance?

Eventually the Los Alamos scientists, with a little help from Edward Teller, figured out what was happening and were able to explain the reason for that curious track. They had discovered cold fusion—a totally new process in which the incoming muons were acting to catalyze the fusion of a light hydrogen nucleus (a single proton) and a deuterium nucleus (a proton and neutron). Somehow the addition of the muons was helping the two nuclei to come close enough for quantum tunneling to do the trick and allow them to fuse—even at the low temperatures of liquid hydrogen! The mysterious 1.7 cm path was the direct and characteristic signature that cold fusion was taking place.

When Alvarez and his co-workers got down to doing some historical homework, they discovered that what they had observed had in fact been predicted some years earlier by F. C. Frank and by Andrei Sakharov (the same Sakharov who had designed the Soviet H-bomb, had come up with the theory of the tokamak, and was later to win a Nobel Peace Prize for his work on human rights). Frank and Sakharov had independently argued that when an elementary particle called a muon is hanging around, then the fusion of a deuterium and a hydrogen nucleus is much more likely to take place. Muons were allowing the "cold" nuclei to approach closely. Thus, muons were the missing ingredients that made cold fusion a reality.

To understand how a muon could play such an amazing role, it is necessary to leave the world of nuclear physics and enter a new domain: the world of the chemist's molecule.

THE HYDROGEN MOLECULE

Hydrogen gas is the same gas that filled the prewar airships like the famous *Hindenburg*. It is the lightest gas known and burns explosively when mixed with the right amount of oxygen to form molecules of water (H_2O). Hydrogen gas is composed of hydrogen molecules, H_2. Remember that, as shown in the previous chapter, the lone electron on a hydrogen atom wants to pair up. The easiest way to do this is for two atoms to enter into partnership and share their respective electrons as H_2.

In a hydrogen molecule, two hydrogen nuclei are held together some $1/_{7,000,000}$ cm apart. On our scale of things, this may seem very close, but viewed from nuclear dimensions, the distance is immense. At that distance, a nuclear reaction is out of the question.

But what about quantum tunneling? Isn't there some finite probability that one of these nuclei in the hydrogen molecule may find itself close enough to the other to fuse? The chances are incredibly slim. In fact, you would have to wait many, many times more than the age of the universe for this to happen. Universes upon universes would pass away, and still no fusion would have happened. On the nuclear scale of things, a hydrogen molecule is totally stable. Thus, quantum tunneling allows the nuclei in a hydrogen molecule to fuse, but the time scale involved is astronomically long.

But nature has another card to play. This card involves a new and bizarre hydrogen molecule, a molecule so tightly compressed that, within such a bizarre form of hydrogen, the probability of quantum tunneling is far higher.

The first chapter pointed out that atoms consist of nuclei and electrons. Now it turns out that the electron has a very curious twin called a muon. For a reason that is not understood at present (see Peat, *Superstrings and the Search for the Theory of Everything*), nature appears to replicate itself by producing not just one type of electron. The electron's twin, the muon, is identical in every way except for its weight, which is 200 times greater. (There is also a third type of electron called a tau, but it need not concern us in this book.)

If the muon is in every way identical to the electron (except for its much greater weight), then why couldn't there be a new type of atom, a muon atom with a single muon revolving around the nucleus in place of a single electron? In fact, scientists have been able to make "muonium," as they call it—a hydrogen atom seen through the wrong end of a telescope.

Figure 3-6

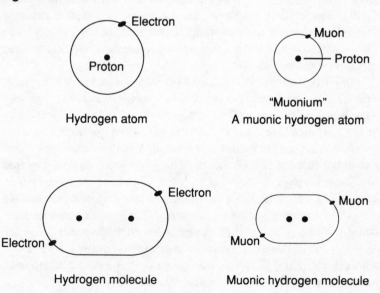

This muon twin of a regular hydrogen atom is similar in every way except that its whole scale is different. It consists of a negatively charged muon revolving around a nucleus. But because the muon is 200 times heavier than the electron, it must "orbit" much closer to the hydrogen nucleus, 200 times closer.

And if there can be a muonic hydrogen atom, why not a muonic hydrogen molecule? In fact, this too is a scientific possibility. If two muonic hydrogen atoms were to meet, then they would want to share their muons and would form an H_2 molecule in a scaled-down form. The two nuclei in a muonic hydrogen molecule would be much closer together, and being close together implies a greater probability of quantum tunneling and therefore nuclear fusion. Would it be possible to use muons to bring about cold nuclear fusion?

This was the idea that Sakharov and Frank had proposed through purely theoretical arguments. This was the same phenomenon that Alvarez and his colleagues saw several years later when examining the tracks of elementary particles. The essence of muon-catalyzed cold fusion is that the muon acts to bring the nuclei close together, close enough to allow quantum tunneling. The muon effect compresses the hydrogen nuclei, but without having to use external force. It is like the ancient art of judo, for it uses the natural inclination of hydrogen nuclei in order to bring about their annihilation. Muon fusion is a gentle process, unlike the violence of high temperatures and magnetic bottles.

The discovery of muons made it possible to completely explain the Los Alamos particle tracks. In the first step, a muon enters a chamber containing hydrogen (proton nucleus) and deuterium (proton-plus-neutron nucleus). This is clearly seen on the photographic track, but then this track vanishes only to appear later.

Helped by Edward Teller, the group determined that as the muon moves through the chamber, it forms a scaled-down hydrogen atom with a nucleus of deuterium. This initial step results in an atom with a single muon outside a nucleus containing one proton and one neutron. The next step is the formation of a molecule. Since most hydrogen nuclei are light rather than heavy, the chances are the result will be a molecule of HD, containing one deuterium and one light hydrogen nucleus.

Figure 3-7

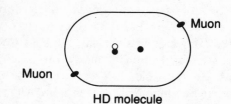

HD molecule

But now the deuterium nucleus and the proton are brought close together and held together long enough for quantum tunneling. The result is fusion, the release of energy, leaving behind a nucleus of helium 3 (an isotope of helium that contains two protons but only one neutron):

$$p + d = {}^3He + Energy$$

Figure 3–7A

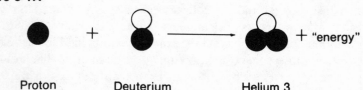

Proton   Deuterium   Helium 3

What is particularly exciting is that the muon itself is thrown out in the fusion reaction. It is now free to encounter another nucleus and give rise to another scaled-down hydrogen atom and another fusion reaction. Again this muon is ejected from the fusion reaction with the same characteristic energy. (In physicists' terms of measurement, this energy is 5.4 MeV.) This amount of energy is roughly similar to that given out when two hydrogen nuclei fuse.

It turns out that the length of a track on a photographic plate is exactly related to the energy of the particle. Physicists knew that a muon with 5.4 MeV will travel for 1.7 cm—exactly the length of the track discovered by Alvarez. In other words, the curious pattern of tracks that had been discovered by accident clearly indicated that cold fusion was taking place.

Since the muon is released after each fusion, it can go on to assist in more and more fusions. In fact, the whole process is called muon-catalyzed fusion. As soon as Alvarez and his colleagues

Figure 3-8

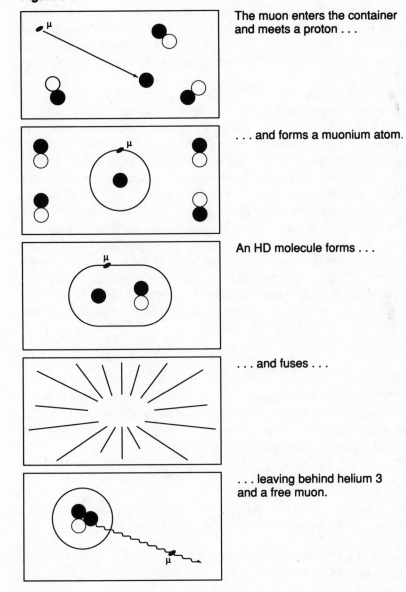

The muon enters the container and meets a proton . . .

. . . and forms a muonium atom.

An HD molecule forms . . .

. . . and fuses . . .

. . . leaving behind helium 3 and a free muon.

realized what was happening, they thought they had cracked the age-old problem of energy. Now fusion energy could be harnessed, not at tens of millions of degrees but at room temperature. All that was needed was a ready supply of muons, with each muon producing hundreds and hundreds of individual fusions. It was a period of great elation.

But when the whole process was worked out in detail, it did not look so attractive any more. Maybe cold fusion was not going to save the world. The problem is that an elementary-particle reactor is needed to produce the muon in the first place, and this requires a great deal of energy to run—only a small percentage of which goes to make muons. When the energy costs of producing muons and running a cold-fusion reactor are added up, they do not look attractive. Certainly energy is produced in the cold-fusion reaction, but not enough to balance the energy drain of the particle accelerator.

Muon fusion works on the margin; it is just the wrong side of creating a world with an endless supply of energy. It was left to Steven Jones of Brigham Young University to struggle with the problem of muon fusion, trying again and again to tip the odds in favor of the human race. It was one of these attempts that resulted in all the excitement of March 1989.

FUSION REACTIONS

The muon-catalyzed fusion of a nucleus of light hydrogen (a proton) and one of deuterium (neutron plus proton) is not the only possible fusion reaction that can be helped by the presence of a muon. It should also be possible to fuse two deuterium nuclei:

$$d + d = dd = {}^3He + n$$

or:

$$d + d = dd = t + p$$

Another approach is to fuse a nucleus of deuterium and one of tritium (the reaction favored by the high-temperature fusion groups):

$$d + t = dt = {}^4He + n$$

Figure 3-9

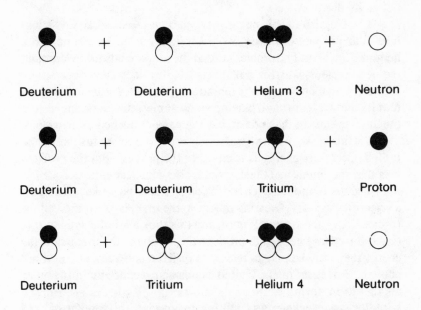

Deuterium   Deuterium   Helium 3   Neutron

Deuterium   Deuterium   Tritium   Proton

Deuterium   Tritium   Helium 4   Neutron

Physicists have worked out the various probabilities for each reaction and have shown how each of these would be enhanced by the presence of a muon, the muon forcing the two nuclei much closer together. On the face of it, the picture should be rosy. In each case, the muon is set free by the fusion reaction itself and should go on to act as midwife in the birth of yet another fusion. Admittedly it costs a lot of energy to create a muon in the first place, but if that one muon can be put to work again and again, it should pay for itself.

But what nature gave with one hand it took away with the other. A muon may be able to bring two nuclei close together to fuse, but it does not live forever. In fact, the muon decays after only one second of life.

In that second, the muon must locate a nucleus, form an atom, bond this atom with a second atom to form a molecule, hold the two nuclei close together, and wait while the fusion reaction takes place. Then, set free, it is able to repeat the process again and again. One second may seem a small time to us, but at the atomic level it is a lifetime. The muon should catalyze many, many fusion reactions before its death.

The blemish in this beautiful design was what is called "sticking." The product of fusing two deuterium nuclei, for example, is a helium 3 nucleus. The problem is that the muon tends to stick around the new nucleus, wasting part of its life in this useless occupation. Like a young teenager who is sent to a convenience store for a carton of milk and hangs around playing video games, the muon spends too much of its time in the undesirable company of the helium 3 nucleus.

Certainly the muon would catalyze a few nuclear fusions before it died. Certainly energy is released in the process. But the problem was that the muon had created far too few nuclear fusions.

The use of muons to create fusion power looked like a dead end, a scientific curiosity. Creating muons in the first place required all the resources of a meson accelerator, and this took a lot of energy to run. Given the energy needed to produce the muons in the first place, the energy they were helping to release in fusion reactions was simply not enough. Scientists in the United States abandoned muon fusion as another lost hope.

But not everyone was willing to give up this dream of cold fusion. In the Soviet Union a number of groups persisted in trying to harness muon fusion. Rather than fusing a nucleus of deuterium with one of light hydrogen, they began to look at the reaction between two deuterium nuclei in the presence of muons. They discovered, for example, that the fusion rate could be increased by raising the temperature. They even worked out some ingenious new ideas about the chemistry of muon molecules that could push the fusion rate even higher. With this reaction, even more energy could be produced before each muon finally decayed.

Then in 1977 two Russians, S. S. Gershtein and L. I. Ponomarev, argued that the reaction between a deuterium nucleus (proton plus neutron) and a tritium nucleus (proton plus two neutrons) would be even more attractive, for in this case the muon could go on to instigate hundreds of fusions before it died. Each time a fusion occurred, the muon was thrown free, where it could assist in a new fusion. The whole process occurred again and again until the muon's life was over.

Two years later, V. M. Bystrisky at the Soviet nuclear research laboratory at Dubna began to experiment on the deuterium-tritium molecule and discovered that the number of fusions could be pushed up by increasing the temperature. Things were beginning to look even more exciting, but the whole experiment had to be abandoned when the Soviet muon accelerator was closed for refurbishing.

The story of cold fusion then moved back to the very laboratory in which it was first observed—Los Alamos. In 1982 a new group began to look at the possibility of muon fusion. This group was composed of Los Alamos scientists, the Idaho National Engineering Laboratory, and physicists from Brigham Young University. Finally Steven Jones had come on the scene.

Jones and his group began working with a muon beam that was directed into a reaction vessel containing deuterium and tritium gases. The vessel, built by the Idaho National Engineering Laboratory, could achieve great pressures, up to 3,000 times that experienced by the atmosphere here on earth and at temperatures from well below freezing to over 500° C. The group discovered that increasing the temperature and pushing up the pressure would cause the fusions to occur faster and faster.

In November 1982 Jones and his colleagues achieved what they called a "scientific breakthrough." They were getting more energy out of the reaction than they were actually putting in. By 1986 Jones was getting out twenty-five times more energy than he put in. And the experiments were not yet optimized, for it should be possible to get even more energy out by increasing both the pressure and the temperature.

But if the Los Alamos experiment could get more energy out of cold fusion than was put in, why didn't this process hit the headlines? The answer was that this was a *scientific* breakthrough, not an engineering breakthrough. Energy is indeed released in the fusion reaction, but his only happens when muons are present. But creating these muons in the first place is very inefficient and requires a separate process involving a lot of energy. When this process is done in a conventional way, more energy is used to run the particle accelerator that creates the actual muons than is ever produced in the fusion reaction vessel.

Figure 3–10   MUON FUSION REACTION VESSEL

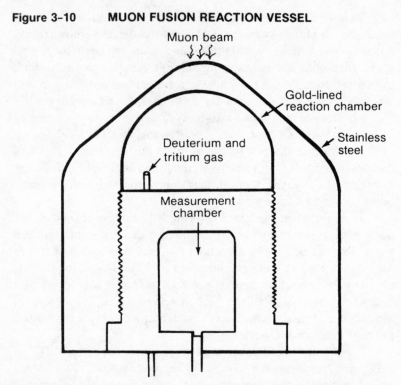

Tritium and deuterium gases are pumped into the gold-lined reaction chamber. A muon beam enters the vessel through the stainless steel wall. Measurements of fusion rate and the effects of temperature are carried out in the lower chamber.

Producing the muons themselves was the stumbling block, the barrier to endless free energy. This was the problem that faced Jones in 1986. Commercial cold fusion seemed only a hair's breadth away. But was there another way around the problem? Was there some process whereby hydrogen nuclei could be forced together, compressed by enormously high pressures and held together while quantum tunneling was allowed to take its course?

Jones attempted to travel along this route by adopting the same sort of technology that is used to create industrial diamonds. Muons had taken him part of the way; brute force pressure was another clue.

But even at that point, he was questioning whether there was some other path, another of nature's secrets that could be used to force hydrogen nuclei even closer together.

It was while puzzling over this problem that Steven Jones hit upon a new solution. The answer, he suggested, lay in forcing the nuclei of deuterium together within a rod of palladium or titanium by means of an electric current. Cold fusion was about to hit the headlines.

Chapter 4
The Utah Alternative

Steven Jones had been pursuing the goal of cold fusion since the early 1980s. His research project had been clearly mapped out, and he had coauthored a popular article entitled "Cold Nuclear Fusion" in the July 1987 edition of *Scientific American*. Working for a number of years on muon-catalyzed fusion, Jones had then begun to look into the possibility of beefing up nuclear fusion through the use of extremely high pressures.

But two other scientists—chemists rather than physicists—had suddenly moved onto the scene and appeared to have scooped the whole cold-fusion story. Just who were Pons and Fleischmann, and where had their breakthrough come from?

FLEISCHMANN AND PONS

Martin Fleischmann was born on March 29, 1927, in Carlsbad, Czechoslovakia, but was brought to England in order to avoid the Nazis. He was educated at Worthing in Sussex and entered Imperial College, London, just after the war, where he received a first-class honors degree. By 1967, at the age of forty, Fleischmann held the Chair in Electrochemistry at the University of Southampton. In that period a professorship at a British university was a far more distinguished achievement than simply being called "professor" at a North American university.

According to Sir Graham Hills, a former colleague, Fleischmann is "a latter-day Leonardo," a man who "would have been brilliant in any subject he had chosen, arts or science." To Professor Ian Fells of the University of Newcastle, "He is a man of great ideas. He has them two a minute. He just hates writing it all up afterwards."

In 1986 Fleischmann received the greatest distinction that can be bestowed by the British scientific establishment—fellowship in the Royal Society. The year before, he had received the Olin-Palladium Medal from the Electrochemical Society. By that time Fleischmann was also holding part-time positions at the University of Utah and at the British nuclear research laboratories at Harwell. With some 250 scientific articles to his credit and a fellowship in the Royal Society, Fleischmann was a highly respected scientist and someone known for the great breadth of his ideas. He was by no means the "obscure chemist" referred to in the May 8, 1989, issue of *Time*.

As far back as the mid-1950s, Martin Fleischmann had been thinking about what happens when an electrical current is passed through a chemical solution. He had been puzzling out the sort of chemical reactions that happen to an electrode dipped into such a solution. (An electrode is a metal rod or strip of metal through which an electrical current enters the chemical solution in an electrolytic cell.) In a very general way, Fleischmann was worrying over the same sort of complex processes that take place inside the battery in your car.

Electrochemistry, as it is called, is a "dirty" topic, for it involves a series of very complicated processes in which many different things are all happening at the same time. In a car battery or a fusion cell, this multiplicity of chemical reactions depends critically upon the physical state of the electrodes, the temperature, and the exact composition of the chemical solution. Just changing a single variable can throw the whole system in a new direction.

Factors such as these were to be critical in the weeks that followed the March 23 announcement. Fleischmann and Pons's critics talked of sloppy work, misinterpreted data, and badly written scientific papers. The Utah researchers replied that if their critics could not observe cold fusion, it was probably because they did not know how to set up a good electrochemical cell!

Electrochemistry requires great skill, experience, and not a little art, all of which Fleischmann had in abundance. The basic idea that led Fleischmann and Pons to cold fusion was born in the 1960s, when the former scientist had noticed some curious data—the appearance of an unusual amount of heat—while working on the electrolysis of deuterium. Since this is not an unusual experiment, other electrochemists must have noticed the heat, but none of them thought to take it seriously. The observation puzzled Fleischmann, who kept it in the back of his mind—possibly there was more in this anomaly than met the eye. Could it have anything to do with a nuclear reaction?

B. Stanley Pons was born in February 1943 and earned his first science degree at Wake Forest University in North Carolina. After beginning research at the University of Michigan and just before getting his Ph.D., he dropped out in order to enter his father's textile business and later to manage a family-owned restaurant in Florida. Although Pons was no longer associated with a university, he continued his interest in scientific research, and he returned to academic life in 1976 when he enrolled in a Ph.D. program at the University of Southampton.

Studying at Southampton brought Pons in contact with his future colleague and friend, Martin Fleischmann. The two men formed the sort of collaboration in which ideas are constantly being generated and bounced from one to the other. Pons earned his Ph.D. in 1978 and went on to take positions at Oakland University and the University of Alberta before becoming first professor and then chairman of the chemistry department at the University of Utah. Martin Fleischmann had been his mentor, but Pons was now able to give him a visiting position in his department at Utah.

Pons is a driven man, a scientist who works long hours in the laboratory and then returns home to work at his computer in the basement. It is said that when Stan Pons reads for relaxation, it is not a novel he picks up but a book of mathematics. His ambition is to publish fifty scientific papers in a single year.

Three-times-married Pons is a somewhat reticent man, not easy with strangers, yet forming very close and generous relationships with just a few friends. His hobbies are skiing, golfing, and fishing. He also collects pocket watches.

Like Fleischmann, Pons published in learned journals and lectured on the topic of electrochemistry. He too had studied the effects of high currents as they passed through chemical cells and the effects of the surfaces of the electrodes used. Independently of Fleischmann, Pons had noticed some curious bursts of heat during electrolysis. It was these first odd results that eventually led the two scientists toward the goal of cold fusion.

Stanley Pons's relationship with Martin Fleischmann is a highly rewarding one. "When Martin is around, there is a special intensity in Stan," says Dr. Richard Steiner of the University of Utah's chemistry department. James Brophy, vice president of research at the University of Utah, has called Fleischmann the creative genius of the two. But others feel that it is a matter of the creative sparks that fly when the two men meet. It was this special relationship that led to Utah fusion. "Without our particular backgrounds, you wouldn't think of the combination of circumstances to get this to work," Pons has said. "We realize we are singularly fortunate in having the combination of knowledge that allowed us to accomplish a fusion reaction in this new way," Fleischmann added.

Making cold fusion work was a matter of getting nuclei close enough and then holding them together while quantum tunneling took place. But all this is easier said than done. For the Utah group, the key insight came while the pair were hiking up Millcreek Canyon near Pons's home. The two scientists had been puzzling about the peculiar properties of palladium. They knew that during electrolysis deuterium atoms are free to enter and move through the palladium metal. Would it be possible to fill the metal with so much deuterium that the nuclei would be driven close enough to fuse?

The two scientists returned to Pons's house and, over a couple of Jack Daniel's, began to puzzle over the details of putting such a cell together. "Stan and I talk often of doing impossible experiments. We each have a good track record of getting them to work," said

Fleischmann. "The stakes were high with this one; we decided we had to try it." Earlier Fleischmann had come up with the idea that if the electrodes were made in the form of rods, then the smaller their radius, the more the lines of electrical force could be made to force atoms together.

The project was worked out in the Pons family kitchen. In fact, the first experiments were fairly simple and done just for fun. "It had a one-in-a-billion chance of working, although it made perfectly good scientific sense," Pons has said. At the time, Pons was running the whole project out of his own pocket. The experiments tended to be run on weekends and were done under the auspices of Pons's own company. For added confidentiality and to save money, Pons was employing his son Joey, who had recently graduated from high school, as an assistant.

While these early experiments appeared to be along the right lines, nothing spectacular happened; in fact, the whole thing seemed to be going nowhere. Then, one night in 1985, the cell they were using grew so hot that it melted down. The two scientists had discussed what would happen if they changed the current in the cell, and that night Pons's son was in charge. He apparently changed the electrical current in the cell, and suddenly the electrode grew so hot that it began to melt. At that moment, Pons realized, the cell must be producing an enormous amount of energy. On that day, cold fusion became a reality for the Utah group.

Pons immediately telephoned his colleague in Southampton with the news, but Fleischmann cautioned that cold fusion was such a hot topic that he had better keep quiet about the discovery. The results would first have to be confirmed and duplicated. Immediately Pons dug deeper into his own pocket and spent some $100,000 of his own money on the cold-fusion project. On that day, the joke was born that while physicists had to go to the government for tens of billions of dollars for hot fusion, chemists could do the whole thing out of their own pocketbooks.

STEVEN JONES

But what of Steven Jones during this period? It may stretch the reader's credulity to believe that two groups could be working in exactly the same scientific field, only fifty miles away, and still not know of each other's existence. Surely, one might ask, they must be stealing or borrowing each other's ideas?

It is certainly true that scientists can be as dishonest as anyone else. They have been known to approach their rivals' graduate students and pump them for ideas. They have been known to use their influence to prevent publication of rivals' results. There have been cases in which a paper, sent for anonymous refereeing, has been deliberately held up by a referee who pushed ahead with the ideas in the paper and published them under his or her own name.

But the history of science is also full of examples where two scientists have forged ahead along parallel tracks without being aware of each other's existence. The most famous example is the letter received by Charles Darwin announcing a theory of evolution remarkably similar to his own. There are many other instances of simultaneous discovery.

It is equally plausible that scientific groups within the same state may be unfamiliar with each other's work, particularly if the two groups belong to different disciplines. It is a common joke among scientists from the same university that the only chance they have to meet and talk together is when they are at an international meeting on the other side of the globe.

So it is not all that surprising that Jones should have been totally unaware of the work of Fleischmann and Pons. After all, Fleischmann and Pons were both chemists, while Jones had approached the problem of cold fusion purely from a physicist's perspective. Jones had first pursued the goal of muon-catalyzed fusion and then in the mid-1980s began to think of using ultra high pressures—piezonuclear fusion—as a way of getting the fusion process to work more effectively.

In March 1986 Jones gave a talk on high-pressure cold fusion, and afterward Paul E. Palmer (now a member of Brigham Young's cold-fusion team) happened to mention that this might be the way the earth generates its own heat. There had always been a great deal of controversy among geologists about the source of the earth's heat. (This topic is discussed in Chapter 6.) Certainly the idea of cold fusion within the earth would solve a lot of problems—and maybe create a few more.

Jones and Palmer began to kick around the idea and went down to San Diego to talk with Harmon Craig, who had been on a geophysical expedition to the South Pacific. During that trip he had collected samples of hot water that rises in plumes from the ocean bed. The helium 3 discovered in these hot-water plumes was later taken by Jones as evidence of cold fusion. Craig agreed that cold fusion within the earth seemed to be a reasonable theory, for geologists at least. Craig also gave Steven Jones information concerning curious observations about the inside of diamonds that could be interpreted as the result of cold fusion.

Back at Brigham Young University, Jones began a brainstorming session. On April 7, 1986, joined by John Rafelski, a theoretician from the University of Arizona, Jones and his colleagues began to play around with a number of ideas. How were they going to duplicate the conditions inside the earth? Presumably cold fusion happened when rocks (which contain around 3 percent water) were sucked down into the earth's mantle along with a variety of minerals. The actual fusion process of the deuterium in the water probably took place inside the hot rocks of the mantle.

The scientists therefore thought about using different metals and minerals. They wondered if cold fusion inside the earth could be triggered by muons. They played around with the idea that fusion could be caused by giving sudden shocks to a metal—for example, with heat or vibration. There were many possibilities to explore, but certainly the best way to get the deuterium into a metal would be by electrolysis.

The major thrust for the group was to set up electrolytic cells in order to pump deuterium into metals. For the cells themselves, Jones chose heavy water laced with the sorts of mixtures of salts that are found in hot springs.

But even if they could simulate geological fusion in the laboratory, the overall effect would probably be very small. For this, Palmer and Jones would need an extremely sensitive neutron detector, since the scientists reasoned that if deuterium nuclei would fuse inside the metal, then they must emit neutrons. It turned out that such a detector was being developed in the physics department by Bart Czirr and Gary Jensen. This would be an ideal way of measuring whether cold fusion was actually taking place.

Where Stanley Pons has been referred to as an ambitious and driven man, the same could not be said of Steven Jones. Jones is a serious scientist who believes strongly in his work and the possibility that it can be used for the betterment of the human race. But he also has another dimension to his life. Born in 1949 Jones was brought up a Mormon, a member of the Church of Jesus Christ of Latter-Day Saints, and subscribes to its code of behavior by dressing in a sober manner and not smoking or drinking. Before beginning his scientific career, Jones served for a time as a missionary in Europe. Family life is important to him; Jones now has six children, and for recreation he enjoys hiking into the hills to camp with his son.

Jones's preoccupation with cold fusion has been partly to come to terms with one of the puzzles of nature, but also to help the human race by providing a new energy source. Jones is polite and softly spoken, at times hesitant and stumbling over his words. Yet a harder edge to his nature shows when he insists on setting the record straight or his version of a particular scientific interpretation. Establishing the truth of his position seems particularly important to Jones, and he sometimes seems impatient and even angry if his point is not coming across.

The Brigham Young and University of Utah groups clearly have a different style. One scientist has observed their very different sources of higher guidance. When Pons and Fleischmann want inspiration, they open a bottle of bourbon and talk together; Steven Jones dedicates his experiment to God.

THE RESEARCHERS' PATHS CROSS

The first Brigham Young fusion cell was built on May 22, 1986, and it at first met with little success. Jones and his team persisted, using different metals and combinations of salts in the "Mother Earth" soup. By the fall of 1988, their results were becoming more encouraging.

On September 20, 1988, the whole picture suddenly shifted when Steven Jones was sent a grant application to review on behalf of the U.S. Department of Energy. The application was authored by Pons and Fleischmann and outlined an experiment to produce cold fusion using an electrolytic cell.

In fact, Pons and Fleischmann had been attempting to keep their work in the dark and not attract attention. But when their results looked more definite, they decided to accelerate their research, and that meant asking for money from the U.S. government. Little did they know what a hornet's nest their application would stir up.

Of the five scientists who studied Pons and Fleischmann's application, four of them were fairly critical. Jones himself stressed that more information would be needed on the project. Pons replied, and a number of telephone calls were exchanged between the two groups, with Steven Jones offering the Utah team his sensitive neutron detector. Finally, on February 23, Pons and Fleischmann met with Steven Jones and Paul Palmer at Brigham Young University.

Pons was quiet during the meeting, complaining of a cold, and left Fleischmann to do all the talking. Later, it appeared that Pons had felt uncomfortable with Jones and, although nothing was said directly, this difference in personality appears to have prevented any further collaboration. Jones was left with the impression that Pons and Fleischmann would visit again, this time bringing one of their cells, to carry out neutron measurements.

Professor Daniel Decker, chairman of the BYU physics department, then joined the group for lunch, where the discussion continued. Jones mentioned that he was about to publish the results of his work and a suggestion was made that the two groups should publish together. However, Fleischmann did not like the idea of sharing his work and insights with another group.

But Jones felt some pressure from Stanley Pons that they should all publish together. Jones reply was to propose the simultaneous publication of independent papers. In fact, the two University of Utah scientists were unhappy with publishing, believing it was too early. They urged Steven Jones to wait. If they could only have six to eighteen months longer, then they would feel more confident of their results. Fleischmann was also concerned that cold fusion had the potential for being used as a weapon, and if their results were presented, he said, then they could be used by the Soviet Union.

Their response presented Jones with a quandary. The American Physical Society was having its big spring meeting in Baltimore on May 1, and the announcements of talks were already being printed in the society's *Bulletin*. Steven Jones had submitted paper J1 3, "Cold

Nuclear Fusion: Recent Results and Open Questions." While the announcement of his lecture was buried in the proceedings, it did contain a bombshell sentence: "We have also accumulated considerable evidence for a new form of cold nuclear fusion which occurs when hydrogen isotopes are loaded into various materials, notably crystalline solids (without muons)."

This was hardly the stuff of which headlines are made, but for those in the field, it would be remarkably suggestive. Copies of the *Bulletin* would soon be mailed out to American Physical Society members, so there was no point in keeping quiet much longer. Jones himself did not feel the necessity to rush any further into print, but if a more definitive announcement had to be made, then why not go back to back, that is, send in independent papers but on the same day. The key would be to find a journal that would publish their results before May 1. (This solution was similar to that suggested by his professional friends to Charles Darwin. They suggested that Darwin and Wallace should submit papers to the same issue of a scientific journal, announcing their independent discoveries of the theory of evolution.)

By now, however, communication between the two groups was beginning to break down. Pons and Fleischmann had not returned to Brigham Young University with their cold-fusion cell. Indeed, the two scientists seemed to feel that beyond the neutron detector, Steven Jones had little to offer them. Jones was clearly going in a different direction and using very different electrolytic cells.

In the end, however, the administrations of the two universities stepped into the gap. On Friday, March 3, the president of the University of Utah called the provost of Brigham Young University and suggested that, in view of the importance of the work, a meeting should be held between the top universities' administrators and chief scientists involved in the projects.

Three days later, the groups met, this time in the presence of the various presidents and vice presidents of the two universities. Again Pons and Fleischmann pleaded to delay publication, and again Jones reminded them of the talk he would be giving at the meeting of the American Physical Society. Finally it was agreed that simultaneous publication was the only answer, and the three scientists later agreed that on March 24 both groups would go to the Federal Express office together and send off their two papers. Until then, nothing was to be said about cold fusion.

Steven Jones says that on March 21 he reconfirmed with Stanley Pons that March 24 was still their date to send off their papers. Imagine, therefore, his shock when, on the following day, he learned that the University of Utah was about to call a press conference to announce the discovery of cold fusion. What had been happening? Why had the University of Utah broken what Steven Jones felt to be a firm agreement on secrecy?

Until early in 1989, Pons and Fleischmann had been pursuing their research in relative secrecy and had paid for it out of their own pockets. But after they asked the Department of Energy for funding, more and more people, including many at Brigham Young University, had begun to learn about what they were doing. Although they wanted to keep quiet about what they were doing, they had applied for patents for their work and were under pressure from the patent lawyers and possibly the university administration—although university officials deny this—to publish their results and thereby strengthen their patent application.

Although the two groups had a firm agreement to publish simultaneously, Fleischmann and Pons decided that on March 11 they would send a "preliminary note" to the *Journal of Electroanalytical Chemistry*, announcing their results. Even though up to March 21 they were in communication with Steven Jones about simultaneous publication, they did not reveal to him that they had already sent off a scientific paper.

Martin Fleischmann has said that during that period the two scientists were under tremendous pressure to go public with their results. The patent lawyers needed to assure the University of Utah's precedence in the subject of cold fusion, and probably the university itself wanted to retain rights to the process. By the weekend of March 17, the two scientists decided to call a press conference, again without informing Steven Jones, and they set the date for March 23, one day before their agreed date for simultaneous submission of papers.

As it turned out, a day before the press conference, the *Financial Times* of London ran an article on cold fusion, and reporters began to telephone the U.S. Department of Energy and the two universities. On the same day, Brigham Young University also learned "with shock and dismay" about the impending press conference. Professor Grant Mason, dean of Brigham Young's College of Physical and Mathematical Sciences, rang Dr. James Brophy, vice president for research at the University of Utah, and told him that if the press conference were held, this would be taken as a violation of the agreement between the two universities.

In fact, there was still a possibility that this press conference would be abandoned. At ten o'clock the night before, Pons and Fleischmann came in to check out the data on a forty-eight-hour run. Pons told Dr. Richard Steiner, associate chairman of the department, "If the results don't continue the way they've been coming in, I'll call the whole thing off, because I'm not going to announce unless the experiments continue to point in the right direction."

But the press conference was held, and on that day, the world at large first heard the term *cold fusion*. On the same day, Steven Jones faxed his own announcement of nuclear fusion in an electrolytic cell to the journal *Nature*. Since the paper was faxed on the evening of the 23rd it was logged by *Nature* as having been received on March 24th.

On the following day, Fleischmann and Pons turned up at the Federal Express office at the prearranged time with a copy of their paper for *Nature*. Jones, it appeared, did not want to appear in front of the TV cameras at the airport, where the Federal Express office is located. After hanging around waiting for Jones, they eventually telephoned Brigham Young University to ask what had happened to him. They learned that Jones had already sent off his own paper and left for a vacation with his family, believing that all agreements between the two groups were void. At this point, each group felt that it had been deceived by the other and that there could never be peace between them.

Although some weeks later Fleischmann was to apologize to Jones for calling the press conference and both men were to shake hands for Italian photographers, things had already gone too far. On April 2, for example, the *Deseret News*, an evening newspaper published in Salt Lake City, carried an article insinuating that Jones had pirated ideas from the Fleischmann and Pons grant application he had reviewed.

According to *The Scientist*, a newspaper for professional scientists, Stanley Pons believes that Jones began to work on the chemical aspects of cold fusion after reading the University of Utah grant application. "In all my scientific life," Pons is quoted as saying, "I have never seen a situation where a proposal was sent to a certain person, who calls up and says, 'Tell me more,' and who then immediately reveals himself as the reviewer and suggests collaboration. I had no idea when he was going to go public."

In fact, Jones's interest in simulating conditions inside the earth through electrolysis was stimulated by Paul E. Palmer before Jones had ever heard of Pons and Fleischmann's research and represented yet another step in his long-term search for a way of fusing hydrogen nuclei at room temperature. Nevertheless, rumors continued to circulate, even though it must have been clear to any scientist involved that the electrolytic cells Jones was using were totally different from those of Pons and Fleischmann. For example, Jones used a chemical "soup" that would have totally poisoned a Pons and Fleischmann cell. Indeed, the two groups were looking at very different phenomena.

By June of 1989, speaking to the author of this book, Jones said that Martin Fleischmann was "a good guy, a really good guy" and that he felt that the tension from the University of Utah may be softening. Even so, it would be a long time before some very painful wounds would be healed.

THE UTAH COLD-FUSION CELL

As the weeks went by, rumors grew and fusion fever heated up. But what exactly had Fleischmann and Pons achieved in that tiny cell of theirs that high-temperature-fusion physicists had been pursuing for decades, spending tens of billions of dollars?

The Pons and Fleischmann cell contains two key ingredients—the metal palladium and an electrical current that is passed through heavy water and into this metal. It is what chemists call an electrolytic cell, and within it deuterium atoms are driven into the palladium at such concentrations that they begin to react and fuse.

Understanding the construction of a Fleischmann and Pons cell begins with that curious and rare metal, palladium. In the earth's crust, it is half a million times rarer than iron. Palladium occurs in natural alloys with platinum and iridium in Colombia, Brazil, the

Soviet Union, and South Africa. In fact, the world's total palladium output is a mere 3.7 million ounces per year, with 95 percent of this coming from the Soviet Union and South Africa—a fact that has given rise to concern in the West should cold fusion depend exclusively on this metal.

Following the Fleischmann and Pons announcement, palladium futures shot from $145 per troy ounce to a five-year high of $185 in a matter of days, with the number of contracts sold per day jumping from 350 to over 1,700. A few days later, when speculators had taken their profits, the price dropped to $160.

This gray-white element has been around the scientist's laboratory for almost 200 years and is used in a variety of electrical experiments and as a chemical catalyst. But the most dramatic property of palladium is surely its ability to absorb hydrogen. Under the right conditions, a piece of palladium can absorb up to 900 times its own volume of hydrogen. Hydrogen molecules appear to stick to the surface of the palladium, then split into hydrogen atoms, which burrow into the metal, sliding easily between the metal atoms.

This property of hydrogen absorption has been known for a long time. Scientists regularly use palladium as a way of storing hydrogen isotopes in certain of their experiments. At first the hydrogen atoms simply slide between the palladium atoms, but as the metal becomes filled with hydrogen, its structure actually changes.

It was in thinking about this curious property of palladium that Pons and Fleischmann got their first ideas about cold fusion. As the hydrogen atoms enter the palladium, they give up their outer electron to move around with the other electrons of the metal itself. The hydrogen nuclei then begin to pack together. Would this packing, Fleischmann and Pons asked, be tight enough to produce fusion?

Since the business of pumping hydrogen gas into palladium had been going on in laboratories for decades and no one had yet spoken about nuclear fusion, Pons and Fleischmann guessed that this effect, by itself, was not enough. (But as we shall see in the next chapter, some historical research later revealed that two German chemists, Fritz Paneth and Kurt Peters, claimed to have detected nuclear fusion in palladium as far back as 1926.)

What was needed was some additional force to set the deuterium nuclei even closer together. Pons and Fleischmann realized that the most practical way of forcing hydrogen atoms into palladium metal would be to use an electrolytic cell.

Electrolysis

In moving from Jones's muon-catalyzed fusion reactions and high-pressure diamond-anvils to electrolytic fusion cells, we are leaving the relatively clean and well-understood world of physics for the complex and confusing universe of chemistry. There is considerable rivalry between chemists and physicists, for the chemists are sometimes looked down upon by their rivals as engaging in "bucket experiments." While the physicist can work in a well-controlled environment on a single well-defined system, the chemist is forced to work much closer to the nitty-gritty face of nature and must be content with all its idiosyncrasies and complexities. While physics works from "first principles" and concerns itself with the basic forces and particles of nature, chemistry is much closer to everyday life.

Chemistry is concerned with unraveling what goes on in a fire, in an oven during baking, when beer is brewed, and in the thousands upon thousands of processes that happen in the human body. In terms of your automobile alone, chemistry is involved in the science of the paint drying, rust on the body, the chrome on the bumper, the rubber in the tires, the gasoline in the tank, the plastics of the seat, the oils in the engine, the coolant in the radiator, and—to return to our subject—the battery that supplies electrical power and recharges as you drive. All this is chemistry. It's complex and often difficult to understand, for many different things are happening all at once, and all depend very critically on such things as temperatures, concentrations, the physical state of a surface, or the presence of a tiny quantity of a catalyst.

In fact, the profound difference between the subjects, and the methods, of chemistry and physics explains some of the hot air that has been expended on the topic of cold fusion. When Fleischmann and Pons made their announcement, it looked as if chemists were trying to tell physicists their business. But as confirmations and refutations followed, it sometimes seemed the physicists were accusing the chemists of not knowing how to detect a neutron or properly measure a gamma ray, to which the chemists would respond that the physicists did not understand how to run an electrochemical cell.

Electrochemistry is a particularly useful subbranch of that general science called chemistry. Its applications are responsible not only for the battery in your car, but also for the power in your digital watch and transistor radio. Electrochemistry is also involved in putting the chrome on your car bumper and the gold and silver plating on cutlery and a variety of other specialized surfaces. In electrochemistry an electrical current is passed through a chemical solution from one electrode (metal rod) to another. This is exactly how Pons and Fleischmann planned to create their cold-fusion cell.

In fact, the Fleischmann and Pons cell is a simple variant on an experiment that is performed in many school chemistry departments and at science fairs. Water (H_2O) can be broken into its component atoms, hydrogen and oxygen, by electrolysis. An electrical current is passed from one electrode to the other through the water itself.

Figure 4-1

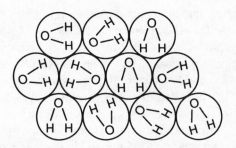

Ordinary water composed of H_2O molecules

This current encourages hydrogen and oxygen atoms in the water molecules to separate. In doing so, the hydrogen atoms leave behind their electron—each hydrogen atom becomes a positively charged nucleus. Oxygen atoms in the water gain two additional electrons, becoming negatively charged.

Each positively charged hydrogen nucleus is attracted toward the negative electrode. Once it reaches and sticks to the surface of the electrode, it is able to pick up a spare electron and turn back into a neutral hydrogen atom. Now the hydrogen atoms on the surface of the

Figure 4-2   THE ELECTROLYSIS OF WATER

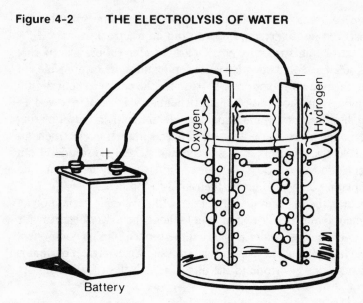

electrode combine into hydrogen molecules (H_2), which bubble off as hydrogen gas. At the other, positive electrode, bubbles of oxygen gas form. The overall effect of electrolysis of ordinary water is that hydrogen bubbles up at one electrode and oxygen at the other.

Fleischmann and Pons were attempting a variation on this basic experiment. They intended to pass a current through heavy water so that, instead of the deuterium gas bubbling off at the negative electrode, the individual deuterium atoms would be driven into the metal itself. This is where palladium and its curious properties came in. If the conditions were exactly right, the deuterium atoms would not recombine into D_2 molecules and bubble off, like the oxygen at the other electrode, but would enter the palladium itself.

Driving the cell for hour after hour would pump ever more deuterium into the electrode. Pushing up the voltage on the cell itself increased the "electrical pressure" on the deuterium nuclei and crowded them even more tightly into the electrode. In a sense, it was as if a great pressure were being created, far greater than could ever be achieved by the "diamond anvil" used by Jones. In fact, Pons and Fleischmann estimated that, with the voltage they were using, they

Figure 4-3

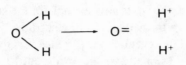

The water molecule dissociates into electrically charged oxygen and hydrogen atoms.

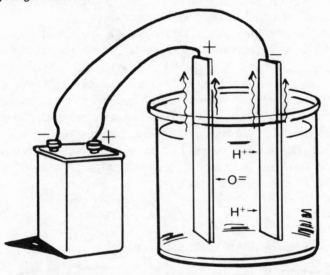

Electrically charged atoms are attracted toward the electrodes.

were achieving a pressure equivalent to 100 trillion trillion (10^{26}) times that of atmospheric. The conditions inside the palladium rod, they claimed, would be equivalent to those inside the core of a star!

Getting these conditions just right, however, was no simple matter—a fact that Pons and Fleischmann were to use to rebut their critics who later claimed to see no fusion effect. To begin with, the actual physical condition of the palladium electrode was important. The key was to have the deuterium atoms enter the metal and not bubble off at the surface. For this to happen, the structure of the palladium surface was critical. Being experienced electrochemists, the Utah researchers learned how to prepare the electrode and ensure that it would remain active.

With the fusion cell correctly prepared, the next step was to allow the electrical current to pass long enough for the concentration of deuterium to build up in the palladium electrode. This dormant period of buildup lasted for several days. The idea was that deuterium had to be pumped into the negative electrode until the deuterium nuclei were held so close that quantum tunneling would allow them to fuse. At this point, heat—a great deal of heat—would be released in the electrode and cold fusion would have begun.

The procedure may sound simple enough, but it took Fleischmann and Pons several years of experimentation and trial and error to get it right.

But how would they know when fusion was taking place? There are several characteristic "signatures" that indicate a nuclear reaction is going on. The Utah scientists were attempting to detect all of them, but for Fleischmann and Pons the most convincing argument was the heat being released. The degree of heat in the palladium electrode suggested a release of energy that was far higher than anything achievable in any chemical reaction.

Figure 4-4

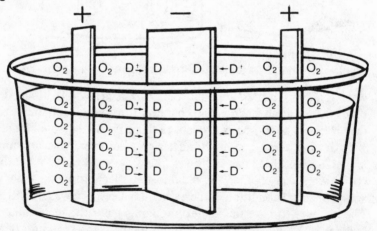

In the Fleischmann and Pons cell an electrical current is passed through heavy water D_2O. While oxygen gas bubbles off at the positive electrodes, deuterium nuclei enter the palladium metal of the negative electrode.

This point is important, because the claim made by Pons and Fleischmann is very different from that made by Steven Jones. While Jones argued that nuclear fusions were occurring at a very low rate, far too low to produce any measurable heat, Pons and Fleischmann claimed that a very large amount of excess heat was being generated in their cell. They firmly concluded that nuclear fusion was occurring. Later, the physicists who began to criticize these results found unbelievable the enormous amount of energy Fleischmann and Pons claimed was being given out by their fusion cell.

What sort of fusion reaction was going on in the palladium electrode? Conventional nuclear physics suggested:

Deuterium + Deuterium = Tritium + Light Hydrogen + Energy   (1)

and:

Deuterium + Deuterium = Helium 3 + Neutron + Energy   (2)

Fleischmann and Pons argued that carrying out the reaction inside the palladium should make possible another sort of fusion reaction:

Deuterium + Deuterium = Helium 4 + Energy   (3)

While reactions (1) and (2) generate nuclear radiation in the form of neutrons and gamma rays, reaction (3) is a "radiationless" form of energy production. It promises a totally clean form of nuclear energy with no need to worry about radioactivity.

Later, two chemists at the University of Utah, Cheves Walling and Jack Simons, were to develop a theory to explain this process. They suggested that a form of nuclear fusion was occurring in which the energy released is given directly to the metal so that no nuclear radiation or neutrons are produced. In short, they claimed that a totally clean form of nuclear energy was theoretically possible.

In fact, this theory was to complicate matters even further and lead to an additional split between the followers of Utah fusion and those of Brigham Young fusion. According to conventional nuclear physics, with each act of fusion, some nuclear radiation should be released. To determine the amount of fusion occurring, one would simply measure the amount of radiation present. More radiation means more nuclear fusions, and more fusions mean more heat. There should be a simple, direct relationship between the intensity of the

Figure 4-5

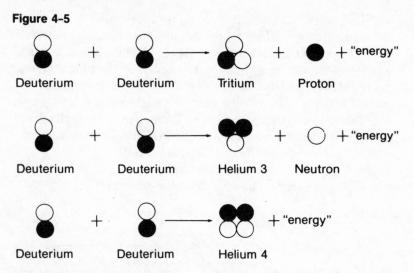

nuclear radiation and the amount of heat generated in the cell.

The problem with this argument was that the large amount of heat being generated in the Pons and Fleischmann cell should be accompanied by enough radiation to fry the scientists alive. But this simply was not happening—which caused some scientists to scratch their heads in disbelief. There was plenty of heat but not enough nuclear radiation.

According to Robert L. McCrory of the University of Rochester's Laboratory of Laser Energetics, for example, if nuclear fusion was really taking place, then the only way to make sense of all that heat was to have a trillion neutrons being emitted each second—enough to kill everyone in the room!

By now the following joke had begun to circulate around the world's laboratories:

FIRST SCIENTIST: Have you heard about the dead-graduate-student problem?
SECOND SCIENTIST: No, what's that?
FIRST SCIENTIST: There are no dead graduate students.

The missing dead graduate students were the key problem in interpreting Fleischmann and Pons's results. With all that energy being given out, there must be an enormous flux of neutrons, enough to kill all the graduate students in the University of Utah's chemistry building. Since no dead graduate students were detected, there were no excessive bursts of neutrons. The answer was to propose some new, nonradiative form of nuclear fusion.

The conclusion, therefore, was either that Fleischmann and Pons were wrong or that some new and exotic form of nuclear fusion was occurring.

If the exotic new fusion reaction (3) proposed by Walling and Simons were the only reaction in town, then everything would be simple. There would be pure heat and no radiation. But the Utah group estimated that one in every 10 million fusions would still go along the conventional path—reaction (2)—and that meant some nuclear radiation. The rest of the fusions, the overwhelming majority of them, simply released their energy with no accompanying radiation. In short, some but not a great deal of nuclear radiation should be coming out of the cell. Detecting this radiation would therefore be an additional confirmation of nuclear fusion. This in fact is where the controversy began.

According to reactions (1) and (2), the Utah group should see three things:

1. Neutrons

2. Gamma rays—produced as some of the neutrons hit hydrogen molecules in the surrounding water bath

3. Tritium—produced in reaction (1)

Fleischmann and Pons claimed to have seen all three.

Later, as some scientific groups confirmed some of their findings, others were to criticize the way in which the Utah measurements had been carried out and to deny the detection of nuclear radiation. If heat was coming from the Utah cell, the critics said, then it had nothing to do with nuclear fusion.

Neutrons

Neutrons were detected using a monitor designed at the British nuclear research center at Harwell. The detector was originally intended to monitor neutrons for health reasons and was less than ideal. Pons had borrowed the detector from the physics department without specifying what it would be used for.

Since some natural background radioactivity is always present in any room, the researchers first had to measure the strength of this natural background and ensure that any neutrons they detected exceeded this original reading. Their claim was that the number of neutrons coming out of the fusion cell was three times higher than the normal background. In addition, the neutrons were emerging at exactly the energy (2.45 MeV) for reaction (2). Later a number of scientists were to challenge this finding.

Gamma Rays

For experimental reasons, the fusion cell was placed in a bath of water. As neutrons from the fusion reaction hit this water, some of them produce what is called a gamma ray.

Fleischmann and Pons claimed that they had indeed detected gamma radiation, and that it was coming off at the correct energy. Other groups later argued that the gamma ray was not coming from the fusion cell at all, but from natural radioactive contamination in the laboratory.

Tritium

The by-product of reaction (1) is tritium, the heaviest isotope of hydrogen. The Utah group claimed to have detected the presence of tritium in the fusion cell. This finding also was to be disputed.

The detection of all three characteristic signatures of a nuclear fusion, along with such a great release of energy in the form of heat, was sufficient to convince the Utah scientists that nuclear fusion was going on in their glass cells.

Convinced that they had overwhelming evidence of nuclear fusion, Fleischmann and Pons wrote up their results for the scientific world. Their formal scientific announcement appears to have been written in two forms: a "preliminary note" submitted to the *Journal of Electroanalytical Chemistry* on March 13 and a paper to *Nature* that was later withdrawn when the journal pressed the scientists for more data.

But long before "Electrochemically Induced Nuclear Fusion of Deuterium" appeared in the *Journal of Electroanalytical Chemistry*, it had been flashed around the world on fax machines and computer networks. Everyone knew what Fleischmann and Pons had been up to—although many scientists were highly critical of the lack of detail given in their paper. Some pointed out that their preliminary note did not have to meet the same rigorous scientific scrutiny as a regularly refereed paper; others noticed that one of the editors of the journal was a friend of Martin Fleischmann's.

When the published paper finally appeared, it contained an erratum, adding the name of Hawkins, their graduate student, to those of the two other scientists. This naturally gave rise to even more rumors. Had the university administration urged Fleischmann and Pons to keep the student's name off the paper in an effort to keep down the number of people involved in a patent claim?

THE BRIGHAM YOUNG ALTERNATIVE

Within a few days of the University of Utah press conference, the results from Brigham Young were also being flashed across the globe. "Observation of Cold Nuclear Fusion in Condensed Matter" by Jones and his co-workers was received by *Nature* on March 24, formally accepted on April 14, and published on April 27—but by that time most scientists had already received their own preprinted copies of the paper by fax and computer network. The paper bears the names S. E. Jones, E. P. Palmer, J. B. Czirr, D. L. Decker, G. L. Jensen, J. M. Thorne, S. F. Taylor, and J. Rafelski, the last affiliated with the University of Arizona. It also acknowledges the valuable contributions of several other scientists.

The problem in comparing the Brigham Young results with those from Salt Lake City was that, although both claimed to be dealing with cold nuclear fusion, the results of Jones and his co-workers were very different from those of Fleischmann, Pons, and Hawkins. Most significant was the absence of excessive heat in the Brigham Young cell. Jones had certainly observed nuclear fusion, but if any heat was released, it was too small to measure. Yet an enormous heat release was the central point in the University of Utah experiment.

What was going on? What was the value of cold fusion if it did not produce abundant energy?

The details of the Brigham Young cell are given in Appendix 3. In essence, the group was working along the same general track as the scientists at the University of Utah, but the composition of the Brigham Young group's cell was quite different. As the University of Utah team had, Jones and his colleagues used an electric current to drive deuterium into the negative electrode of a cell. But in Jones's case, the chemicals in the cell were far more complicated—a mixture of salts of iron, nickel, palladium, calcium, lithium, sodium, gold, and titanium was added to the heavy water. Later Jones was to refer to this "Mother Earth" mixture as "typical of volcanic hot springs."

In fact, the table next to the group's neutron detector is filled with chemical reagent bottles, each one containing a different salt or metal that was tried out in various experiments. In addition to using palladium as the electrode, the group also reported work using the much cheaper metal titanium (in its "fused" form) and had even experimented with iron, nickel, copper, zirconium, lanthanum, titanium, and lithium-aluminum. The Brigham Young group took particular care with the surface of the electrodes, using palladium in thick foils and its "mossy form," as well as cleaning and roughening the surfaces.

While Fleischmann and Pons had run their cell for around one hundred hours before detecting the excess heat of fusion, the Brigham Young group detected fusion after only one hour but found that the process began to fall off after eight. In essence, the magical mixture of salts they had added to the cell was accumulating on the electrode and "poisoning" it.

The two papers were worlds apart. At Utah intense heat was released after the experiment had been running for one hundred hours. At Brigham Young the whole reaction was over in eight hours. The Utah cell was projected as the prototype of commercial energy cells of the future. At Brigham Young, nuclear fusion occurred at a very low rate, and any heat produced was too small to be noticed. While the Utah scientists were willing to project energies of tens of kilowatts per cubic centimeter of electrode, the Brigham Young group referred to a small effect that, nonetheless, "opens the possibility, at least, of a new path to fusion energy."

Over the next weeks, Steven Jones would emphatically deny that the heat seen by Pons and Fleischmann could have anything to do with nuclear fusion.

Jones and his group did not measure heat or gamma rays. Their key piece of evidence was the appearance of neutrons. They assumed that they were observing the following reaction:

$$\text{Deuterium} + \text{Deuterium} = \text{Helium 3} + \text{Neutron}$$

and not some other exotic fusion process in which energy is directly given up to the palladium or titanium without any radiation. While Fleischmann, Pons, and Hawkins were claiming that only 1 in 10 million fusions emitted a neutron—the rest occurring without the production of nuclear radiation and giving rise to heat alone—Jones and his group assumed that every neutron detected was direct evidence of a single act of fusion.

Since very few neutrons were actually detected, the Brigham Young scientists were looking at a fusion rate well over a million times smaller than that claimed at the University of Utah. No wonder no excess heat was detected.

Did this mean that Jones and his group lacked the secret of true fusion power? Or could it be that there was a fundamental flaw in the Utah measurements? Was all that excess heat something of an illusion? While Pons and Fleischmann were already planning to scale up their energy production, Jones was talking about commercial cold fusion taking "twenty years to never. If you own an oil well, don't go out and sell it."

The nuclear reactions discovered at Brigham Young could well be a step on the road toward commercial cold fusion, but in their present design, the cells were producing only a handful of neutrons—very few fusion reactions indeed. In fact, Steven Jones had spent the previous years designing a special detector for registering such a low neutron flux—around 200 per hour. That is less than one nuclear fusion each second within the cells.

The two claims were therefore very much at odds: on the one hand a new but very modest form of cold nuclear fusion with only a slight chance of ever blossoming into a commercial possibility, and on the other, immense amounts of heat being produced in a small cell and the possibility of immediate scaling up to much larger energy production.

FUSION FEVER BEGINS

The Fleischmann and Pons press conference shocked the scientific world. Here were two chemists who, with money out of their own pockets, claimed to have achieved the very goal that had frustrated the physics establishment for decades.

The only problem was to determine exactly what they had done. No one seemed to know. As Donald L. Correll, a fusion physicist at Lawrence Livermore National Laboratory, put it, "I still don't have an understanding of what these guys did—and they don't either." Hendrik J. Monkhurst, a physicist at the University of Florida, was content with "Miracles happen."

And what about all that heat coming from the fusion cells? Robert L. McCrory of the University of Rochester's Laboratory of Laser Energetics said, "If I know anything about physics [those results] are wrong."

Within hours of the press announcement, Dan Rather had opened the CBS "Evening News" with a report on "what may be a tremendous scientific advance." The "MacNeil/Lehrer NewsHour" carried excellent coverage of cold fusion. In fact, a group at MIT had videotaped the item. With no other news or scientific information to go on, they crowded around the television screen, running the tape again and again, trying to catch a glimpse of the apparatus and, in particular, attempting to figure out what the neutron counters in the background were registering. Within four days of the Utah announcement, the MIT team had six cold-fusion experiments set up.

Georgia Tech also rushed into the game, but as its laboratory did not have any palladium for the electrode, there was a delay in setting up the cell. Other teams at the Los Alamos and Lawrence Livermore national laboratories and at Texas A&M were building fusion cells. Soon the world would learn about groups working in Italy, Hungary, Japan, Brazil, Argentina, Canada, the Soviet Union, India, Germany, France, Switzerland, and North Korea.

Ontario Hydro, one of the world's major producers of heavy water, was inundated with phone calls from university and industry laboratories. In a single four-day period, it sold 600 kilograms of the substance at $350 a kilogram!

Martin Fleischmann was himself a consultant to Harwell, the research laboratories of the United Kingdom Atomic Energy Authority, and appeared to be helping their scientists set up a fusion experiment. But Dr. Ronald Bullough, chief scientist at the laboratory, appeared to be skeptical about the Utah results. The chairman of the authority told a British Commons Energy Committee that he did not see cold fusion as having any important role to play in the first half of the next century. It appeared that Fleischmann left the Harwell groups in a fit of pique when they could not agree on how to set up the experiment. Later Harwell failed to observe the Pons and Fleischmann effect.

At the University of Washington, Seattle, two graduate students, Van L. Eden and Wei Liw, soon claimed to have seen small quantities of tritium produced in the reaction. "We've been going at it twenty-four hours a day for ten days," Eden said. "We look back at our notebooks and say, 'God, did we do all that? It's crazy.'" After running the experiment for three hours, the students detected tritium. As a control, they repeated the experiment using ordinary water but could detect no tritium under those conditions.

All that these groups had to go on at the beginning were newspaper reports. The normal process of science had been totally bypassed, for chemists and physicists could only rely upon what was coming in over the wire services and what they had seen on television. Science had explosively entered the electronic age. The telephone, fax machine, and computer network were taking the place of the seminar, conference, and learned journal.

The best that scientists could do was to swap guesses, ideas, and intuitions over computer networks like Arpanet and Bitnet. John Sheffield, director of the Oak Ridge National Laboratory, felt that scientists were all in the position of a cook in the Middle Ages who had been told legends of a mysterious new dish called a soufflé. The only recipe to go on was that it contains milk, eggs, flour, and a bit of cheese. Even knowing the ingredients was not enough; the key was to know the right way of putting them together.

Not everyone was happy with this new way of working. The University of Toronto chemist and Nobel Prize winner John Polyani felt that the "due process" of science was being circumvented. "One cannot settle these very subtle questions of interpretation of experimental findings by the scientific community addressing each other through the press. I think there is a danger to science and also a danger to the relationship between science and the public if we continue to enthuse people and to raise expectations that are based on ephemeral bits of evidence day to day. Then we do bring the subject of science into disrepute."

Just over a week later, as some of the first independent results were beginning to trickle in, Jones and Fleischmann were both giving seminars. The University of Utah scientists were also pushing for funding. After all, the money spent on their experiments appeared negligible when compared to the tens of billions spent on high-temperature fusion. Norman H. Bangerter, Utah's governor, was asking for $5 million "contingent upon confirmation of the experiments" and had called a special session of the state legislature for April 7. There was even a rumor that James Fletcher, former president of the University of Utah and recently retired head of NASA, would become head of a new university center for cold-fusion research.

Meanwhile, at Columbia University in New York, Steven Jones was talking to a room packed with scientists, students, and reporters. He stressed that what he had observed was only a step on the way and that it would be a long time before cold fusion became a commercial possibility. All that the Brigham Young group had detected so far had been neutrons—no heat and no gamma rays. "We have a Cinderella but no shoe," he said.

At a press conference on the same day, Jones appeared to be annoyed at the attention the Utah group was receiving. He contrasted the scholarly care of his own work with what had been going on at "the other Utah university." He produced a page from a laboratory notebook and told a reporter to note the date. The handwritten page contained a sketch of a fusion cell—Jones was staking his claim to have come up with the idea for cold fusion several years ago. He also claimed not to have read the Fleischmann and Pons paper. While other scientists were busy faxing its few pages across the world, Jones pointed out that he had a big pile of papers on his desk and just did not have time to read them all. He even refused a copy of the Utah paper when it was handed to him. Later, Jones told the author of this book that although he had not read the Pons and Fleischmann paper he was sure that he already had a copy in his office.

Meanwhile Martin Fleischmann had turned up at CERN, the European nuclear physics laboratory near Geneva, where he gave a seminar. CERN's director, Carlo Rubbia, winner of the Nobel Prize for his elementary particle discoveries, first ordered all the journalists out of the room. This was going to be a truly scientific meeting. Then Rubbia joked that this was the first time that a chemist had discovered a neutron!

According to a physicist who was present at the time, Fleischmann gave an impressive lecture at which he presented details of his experiment. It was clear from his talk that the British electrochemist believed his Utah results could be explained only on the basis of a nuclear reaction. Such a large quantity of heat could not be produced in any other way. There was also the suggestion that conventional fusion reactions, like the combination of two deuterium atoms to produce helium 3 and the release of a neutron, could not be the answer. What was required was some new and exotic nuclear transformation in which fusion took place without nuclear radiation. Some days later CERN was due to hear from Steven Jones.

On the same day as Fleischmann's talk, March 31, one of the first confirmations came in. Two Hungarians, Gyula Csikai and Tibor Szaricskai at the Lajos Kossuth University in Debrecen, claimed to have duplicated the cold-fusion results. At the Lawrence Livermore National Laboratory, however, the experiment exploded with a loud report, shattering glass all over the laboratory. At least Fleischmann and Pons had warned them.

Just over a week later, on April 10, one of North America's major universities, Texas A&M, had confirmed at least part of the Fleischmann and Pons result. Using a very sensitive calorimeter—a device for measuring the heat given off by a chemical reaction—the university's researchers had detected heat being given off by the reaction, around 60 percent to 80 percent more than they were putting in. But Professor Charles Martin remained cautious: "I would feel a lot more comfortable if we detected fusion, and we have not yet."

If the researchers in Texas had not detected any evidence of fusion, such as neutrons, Georgia Tech was at last up and running and claiming to have detected neutrons but no heat. Cold-Fusion Confusion had just begun. As one scientist put it, "One group sees heat and no fusion, the other sees fusion and no heat. What is going on?"

By now cold fusion was big news, and Martin Fleischmann was reported to have left Britain to continue work in a secret location. "I didn't want everyone sitting on my tail," he said. Fleischmann had been due to turn up on April 12 with his colleague Stanley Pons at the 197th annual meeting of the American Chemical Society in Dallas to give an address to the assembled scientists but at the last moment appeared to have changed his plans. Reporters tracked him down to the Ettore Majorana Science Center at Erice, Sicily, where he was giving a seminar on the conditions of his cold-fusion experiment. From that point on, he would be working at "an undisclosed location."

Back at the American Chemical Society, Stanley Pons faced an audience of some 7,000 when he spoke at a special forum on "Nuclear Fusion in a Test Tube?" Even Elsevier Sequoia, publisher of the *Journal of Electroanalytical Chemistry*, gave a press conference to announce that Fleischmann and Pons's paper would appear on April 6. The scientific editor of the journal was present, and free copies were made available to scientists and reporters.

Pons himself was so mobbed by reporters and scientists that he was forced to change hotels twice and to register under an alias. Back in Utah special security had been established at his laboratory, and on one occasion the scientist's automobile had been blocked in at the airport by a man who demanded "the secret of cold fusion."

At the Dallas meeting, Pons reiterated how the Utah experiment involved passing an electrical current through heavy water and into a palladium electrode. Fusion in the rod produced four watts of energy for every watt used to run the cell. He even produced a picture of a test tube labeled "The Utah Tokamak."

But now Pons was facing tough questions as a number of fusion experts expressed their skepticism and wondered if that release of energy could be provided by some other, nonnuclear process. Harold P. Furth, director of Princeton University's Plasma Physics Laboratory, argued that physicists would not believe it was cold fusion until a certain critical experiment was done.

The major question in the cold-fusion claim was: Could the energy release be explained in some other, chemical way? Cold fusion depends critically upon the fusion of deuterium nuclei in the palladium electrode. The effect could not happen if light water were substituted for heavy water; calculations suggested that quantum tunneling would be unable to fuse light hydrogen nuclei, and no heat would be observed. But since light water is chemically identical to heavy water, if the heating effects are due to some chemical reaction and were not nuclear, then nothing should change if light water were substituted in the cell.

This, in essence, was the critical experiment: run everything as before but substitute light for heavy water. If true cold fusion has been taking place, it will now cease, and no heat will be observed. But if the heat is the result of some unknown chemical effect, then the same heat production will be found with light as with heavy water. This key control experiment would have to be carried out before the scientific world was convinced.

Back in Sicily, Martin Fleischmann was pondering exactly the same topic. Both Utah researchers agreed on the value of such a control. But the problem was that working with light water would destroy the effectiveness of their electrode, which was very difficult to prepare. Later the scientific world was to learn that this control experiment had already been carried out by Steven Jones.

Fleischmann and Jones were to meet face to face during the

Sicily meeting. Earlier there appeared to have been some friction between the two groups. Jones had complained about Utah breaking their agreement to announce cold fusion simultaneously. But now the two men seemed, for the time at least, to have patched up their differences. As far as Steven Jones was concerned, the best thing was to get down to some serious science again.

He would be going to the Italian National Institute of Nuclear Physics, which is built deep in the Gran Sasso Mountain in central Italy. By locating their laboratory underground, the Italian scientists had been able to filter out much of the stray radiation that is created by cosmic rays arriving from outer space. Since background neutrons would be kept at a minimum, it should be possible to obtain more reliable information on the nuclear reactions taking place in the cold-fusion cell. Indeed, the Italian scientists did detect neutrons, but only at the much lower "BYU level."

Jones and Fleischmann also discussed the significance of the differences between their two experiments. The scientific climate, among physicists at least, seemed to be moving toward the more modest Brigham Young approach. Cold fusion was a staggering new concept, but at least Jones and his group were not making such extraordinary claims as to the production of enormous amounts of heat. Fleischmann, for his part, speculated that the differences between their two sets of results could arise from the different scale on which the fusion cells were operating. At Utah the cells had run for over one hundred hours, while at Brigham Young the effect tailed off within eight hours.

As more confirmations of the Fleischmann and Pons effect were reported in early April, other groups were beginning to have serious doubts, and they began talking to their colleagues over the computer networks. As their doubts multiplied, the scientists focused on the American Physical Society meeting of May as an ideal time to release their findings.

By then the news was also out that Fleischmann and Pons's paper had been accepted by the *Journal of Electroanalytical Chemistry* on March 20—three days before that fateful news conference. But what about the papers that both groups had submitted to *Nature*? On April 12, Laura Garwin, one of the journal's editors, announced that the paper by Steven Jones and his group had been officially accepted and would appear in the April 27 issue. Rumors began to circulate as to the fate of the paper by Fleischmann and Pons. It appeared that *Nature* had pressed the authors for more data to back up their claim. Fleischmann and Pons had submitted several extra pages, but these were not considered sufficient. In the end the two researchers had withdrawn their paper.

Several of the groups that immediately tried to duplicate the Pons and Fleischmann experiment ran into trouble by assuming that building a fusion cell would be easy. Others, like Ontario Hydro laboratories, which had entered into a private agreement with Fleischmann and Pons to gain access to their patents firsthand, could read the Utah "cookbook" with its formula for success. Apparently it was crucial to the reaction not simply to drive hydrogen atoms toward the palladium electrode but to have the electrode prepared in exactly the right way so that trace amounts of metals would appear at the electrode at exactly the same time. Groups also played around with the electrodes, using different shapes and sizes as well as different forms of the metal—thick foil and a "mossy" form. Various voltages from 3 to 25 volts were employed, as well as a variety of currents, generally a half to a hundredth of what passes through a normal 100-watt light bulb.

On April 14, Dr. Donald Parker, director of MIT's fusion laboratory, announced that none of the twenty experiments running there had confirmed the Fleischmann and Pons result. It appeared that groups at AT&T's Bell Labs and at the California Institute of Technology also were unable to observe cold fusion. On the other hand, scientists at MIT and the University of California at Berkeley were providing theoretical support for cold fusion and making the results more credible to scientists.

On April 12, the Soviet news agency Tass announced that a group in the physics department at Moscow University had confirmed room-temperature fusion by detecting neutrons in twenty experiments and could "assert with confidence that the nuclear fusion reaction actually takes place." Said Professor Runar Kuzmin, "The experiments are surprisingly simple." Academician Anatoly Logunov, rector of the university, did not doubt the correctness of the experiment. The university was about to embark on a large program of fundamental research into cold fusion, which, Logunov said, is "undoubtedly a very important phenomenon."

Cold fusion was also making a dramatic impact in Japan. That country imports 99 percent of its oil, and its Ministry of Trade and Industry spends more on energy research than on all other technologies combined. At Yokohama National Laboratory one scientist, Dr. Ken-ichiro Ota, began a twenty-four-hour vigil over a fusion cell. The giant Toshiba and Hitachi corporations also got in on the act but were saying little to their competitors. Later however, Hitachi was to announce the confirmation of cold fusion.

In the first week of April, Japanese scientists were comparing notes at a conference in Osaka. Two days later an emergency scientific conference was held at Yokohama National University to a standing-room-only audience.

Japanese government agencies were also involved: the Electrotechnical Laboratory of the Ministry of Trade and Industry, the National Chemical Laboratory, and the Science Agency were now working on cold fusion. By April 4 a group led by Noboru Oyama at Tokyo University of Agriculture and Technology had observed large amounts of heat. The group was about to join forces with the Japanese Atomic Energy Research Institute to confirm whether neutrons were released.

By mid-April cold-fusion reports were coming in thick and fast, and the results were confusing. Georgia Tech, which had earlier reported having detected neutrons, was backing off. "Georgia Tech Eats Crow," ran one headline, when the group admitted misinterpreting its data. Apparently the Georgia Tech neutron detector could be made to register simply by warming it up. "When we put it in hot water, the thing jumped as if there were neutrons," said Professor James Mahaffey. The researchers labeled their original experiment "The Shroud of Turin," but were pressing on with new observations.

By this time, most universities and laboratories in North America had someone either working directly on cold fusion or at least thinking about it. People were asking each other, "Do you believe it? Is it really happening?" A story circulating around MIT in the middle of April had one professor scathingly dismissing cold fusion, saying that the chances of it working were a million to one. A colleague slapped a dollar bill on the table, saying, "Where's your million?"

The showdown was scheduled for May 1, for on that day physicists from all over North America would be assembling in Baltimore for their annual spring meeting. Irrespective of what happened to be on the agenda, the physicists would all be talking about cold fusion.

Chapter 5
Confirmations and Refutations

While scientists all over the world were staring into fusion cells and waiting for neutrons to appear, the administration at the University of Utah was being inundated with telephone calls. A variety of companies wanted to get in on the action and produce fusion cells of their own. These callers ranged from multinational Fortune 500 corporations to a number of small businesses that were attempting to find their own niche in this vast potential market.

In Canada, for example, Electrofuel Manufacturing Company of Toronto manufactures batteries, a field that requires experience in electrochemistry. Dr. Dasgupta's company began work on fusion cells of his own and within a few weeks was applying for U.S. patents. The Toronto company also jumped on the cold-fusion bandwagon by selling a new form of "electrochemical calorimeter"—a very sensitive device designed to measure directly the heat being emitted from a cold-fusion cell.

Ontario Hydro, one of the world's largest producers of heavy water, was getting calls from laboratories, while Marshall Laboratories of Boulder, Colorado, had placed an ingenious advertisement for palladium wire in *Physics Today* with a drawing showing two deuterium nuclei fusing.

Not only were laboratories and corporations getting into the cold-fusion business, but at Texas A&M calls were coming in from investors trying to sniff out the wind of cold fusion as well. At Georgia Tech dealers in precious metals were asking for advice on palladium futures. Venture capitalists also were trying to figure out the best place to put their money.

By mid-April the University of Utah had established a series of teams to develop and exploit the new finding. The interim director of this Cold Fusion Research Project would be Dr. Hugo Rossi, a scientist with experience in the field of elementary particle physics. The university's College of Mines and Earth Sciences was to carry out a series of engineering studies directed toward practical applications. To begin with, the group was going to look at the scaling-up process in which larger fuel cells would be used to create more power. "Possibly the best basis for determining the engineering and technological potential for cold fusion is to demonstrate a small yet complete energy production system is possible and practical," said Dr. Gary M. Sandquist, professor of mechanical engineering.

But a number of questions had to be answered first. "For example, does the reaction, whatever its source, accelerate as the temperature increases, and how can any acceleration be controlled?" asked Dr. Robert F. Huber, professor of electrical engineering. Scientists had dreamed up a number of theories that suggested temperature should play a crucial role in the cold-fusion process. Dr. Robert Boehm, another professor of mechanical engineering, wanted to study a fusion power reactor over the long term to learn how quickly such a reactor could be shut down in an emergency.

There was even the proposal that cold fusion could be adapted to power automobiles and trucks. "This is not very feasible with present-day nuclear reactors," said Dr. Boehm. "Hence estimates of the compactness of a practical unit would be valuable in assessing potential applications."

While Steven Jones at Brigham Young had been working with titanium electrodes, the Utah approach still depended on the more expensive palladium. The metallurgists at the Cold Fusion Research Project were therefore interested in working with cheaper and more abundant minerals. "If research on scale-up becomes a reality, some metal besides palladium will probably prove more effective for making electrodes," said Dr. Sivaraman Guruswamy.

Scientists would also study the effects of nuclear radiation on the electrodes and determine how microscopic flaws in the palladium's atomic structure were influencing the fusion reaction. Other experiments would look at changing the mixture of chemicals that were used in the cells. "Data from studies such as these would put chemical engineers in a good position to develop efficient reactor cell configurations for commercial applications," said Dr. J. D. Seader, a professor of chemical engineering.

By now further confirmations of cold fusion had come in. Scientists at the Indira Gandhi Center for Atomic Research, for example, had observed cold fusion in a titanium electrode. Although they had not measured the increase in heat, they were able to detect neutrons and had also carried out a control experiment in which the heavy water had been replaced by light water. A few days later, Dr. C. V. Sundaram reported a 30 percent increase in neutron emission over natural background levels. Bratislava's Comenius University in Czechoslovakia also claimed to have witnessed fusion.

On the same day, April 17, Pons himself announced that in one cell the fusion reaction had been sustained for 800 hours and was producing eighty times more energy than it consumed. The power output was now up to sixty-seven watts per cubic centimeter of electrode. Nineteen new fuel cells were being set up, and the design for a small-scale power reactor was under way.

At the same conference, Pons was joined by two chemists from his department, Cheves Walling and John P. Simons, who argued that a new type of nuclear reaction was occurring in the electrode, a reaction in which energy went directly into the metal itself and did not emit radiation.

A day later scientists at Stanford University also confirmed cold fusion in a careful set of experiments. In particular they had carried out a control experiment in which they repeated the whole procedure using ordinary light water in place of heavy water. The two cells were contained in a red picnic cooler for two weeks. As predicted by Fleischmann and Pons, since light water is chemically very similar to heavy water but in nuclear terms quite different, the heating effect was not seen with light water. Robert Huggins, professor of materials science, claimed that only when heavy water had been used had the group detected heat "comparable with that reported by Pons and Fleischmann." He also speculated why other groups had not yet confirmed the original experiments: "It sounds simpler than it really is. It's easy for people to do bad experiments."

Two researchers at the University of Florida's Department of Nuclear Engineering Sciences, Glen J. Scoessow and John A. Wethington, claimed to have detected tritium during the cold-fusion process—an important clue that fusion was actually taking place—but not when light water was used. Two groups in Poland, at the Technical University of Gliwice and the University of Wroclaw, claimed to have confirmed cold fusion, as did a group in Brazil.

One of the most striking confirmations for cold fusion came from a laboratory near the town of Frascati in Italy. Francesco Scaramuzzi of the Italian Agency for Nuclear and Alternative Energy (ENEA) had built a variation of the Utah cold-fusion cell. Said Scaramuzzi, "The idea arose from the question 'Is electrolysis really necessary to produce this interaction between the palladium and titanium?' " In fact, it turned out that Scaramuzzi had been thinking about this type of cold fusion for a number of years. The group concluded that an even simpler experiment was possible: rather than driving deuterium into an electrode by electrolysis, the group was pumping deuterium gas into the metal under pressure.

The first experiments began on April 7, and after some failures the group claimed to have measured neutrons 600 times in excess of the normal background. What is more, there appeared to be peaks in this emission of neutrons that coincided with deuterium gas coming out of the metal. But in these experiments no heat was detected.

The group had investigated a variety of metals and chose titanium because of its availability. Professor Umberto Columbo, president of ENEA, said, "I want to point out that this is a great scientific discovery." But he cautioned that it would take years before possible commercial development. The new phenomenon of "Frascati fusion" was hailed by the Italian press as a great national triumph.

Some physicists speculated that the Frascati experiment had been so successful because Martin Fleischmann had been able to help the Italian group before the University of Utah's patent lawyers pulled a blanket of secrecy over the whole process. But if this were true, then why hadn't the Harwell Laboratory in England enjoyed a similar success, since Fleischmann was also their consultant?

As time went on, scientists were to learn of other, bizarre, attempts to instigate cold fusion by bypassing electrolysis. It turned out that in the 1970s Russian scientists had shot lithium pellets saturated with deuterium at a concrete wall and claimed to have detected neutrons. Another idea was to use ultrasound to send a shock wave through the palladium and in this way compress and fuse the deuterium nuclei.

Cold fusion was being taken seriously by a number of legislators. On Wednesday, April 26, Pons, Fleischmann, and Jones were due to testify before the House Science, Space, and Technology Committee. That day was a triumph for Utah fusion as Fleischmann and Pons reiterated their claim and pressed the congressmen and congresswomen for increased funding. Utah University president Chase N. Peterson suggested that a national fusion center be established in Utah. This would take some $100 million to get running, and he suggested that the federal government supply $25 million as seed money.

The Utah group had hired Cassidy and Associates, a lobbying firm, to press Congress for funding. They were accompanied by their business consultant, Ira Magaziner, president of Telesis, Inc., which advised on corporate strategies and economic development for governments. Magaziner pulled no punches: "As I speak to you now," he said, "it is almost midnight in Japan. At this very moment, there are large teams of Japanese scientists in university laboratories trying to verify this new fusion science. Even more significantly, dozens of engineering company laboratories are now working on commercializing it." Magaziner evoked the specter of hordes of Japanese scientists and businesspeople who were about to take over cold fusion.

But there was an alternative, Magaziner said: "It's the alternative that says that America is prepared to fight to win this time." But fighting costs money, and that became the bottom line. "I have come here to ask you, for the sake of my children and all of America's next generation, to have America do it right this time."

Steven Jones and the Brigham Young group urged caution. In their opinion, the heat seen by Pons and Fleischmann could not be the result of nuclear fusion. Although a low level of fusion was certainly occurring in the electrodes, it would be a long time before scientists could arrive at that final goal of commercial cold-fusion power.

The State of Utah was also getting into the energy game, with Governor Norm Bangerter calling a special session on April 7 to request $5 million in research funds. "He that doeth nothing is damned," the governor said, "and I don't want to be damned." By a majority of 96–3 the legislature passed the Fusion/Energy Technology Act, which would provide funds for research, provided that scientific confirmation was met. In addition the act sought to prevent the leaking of information on cold fusion, an aspect that was hotly protested by newspaper reporters.

Not everyone else was happy with the meeting either. "If this fails, we'll have egg on our faces. It'll show the world that Utah is unaware of the proper scientific process," said Merril Cook, a businessman who had unsuccessfully run for governor.

Under the act, the state established a Fusion/Energy Advisory Council, to be appointed by the governor, and charged the University of Utah with setting up an R&D program in cold fusion. By the start of May, half a dozen new companies dealing with cold fusion had been registered, and Fusion Information Center Inc. had been established to offer "a computerized fusion database for scientists and researchers." The company planned to hold a Fusion Impact World Conference in July, in which over 2,000 researchers and business executives would be involved.

Ironically, following the March 23 press conference, Pons and Fleischmann's funding application was finally approved by the Department of Energy. It was this application, reviewed by Steven Jones, that had triggered the whole chain of events leading to the fusion controversy. Now the department was offering $332,000, which the university refused. Norman Brown, director of technology transfer for the university, said, "As soon as we take one dollar of federal money, any discoveries we make become the property of the U.S. government."

The University of Utah, however, was already scaling up its cells to produce far more energy. "This is a natural outgrowth," Pons said. "There are much better designs for getting heat out of the cell and to generate a greater reaction. We'll start construction on that when Fleischmann gets here." Pons was thinking big: "If you can charge very large pieces of palladium, the heat would be considerable. It would be that sort of design that would lead to commercialization. You could set up any size you want. To do that is going to require a lot of additional data and design work."

With four reactor cells already running at Utah, nineteen more were planned for the near future. "The largest will be maybe five times as large, but that will take a long time to charge." With all their experience, the Utah scientists were finding it easier to get the cells to work most of the time. "I think it's clear," said Pons, "that the processing of the palladium is a factor." In fact, Pons claimed that the use of noncast or extruded palladium was the reason that some groups failed to duplicate the experiment.

But not everyone was willing to believe in cold fusion. Michel Barsoum and Roger Doherty from Drexel University, Philadelphia, claimed to have detected heat from a "fusion cell" even when ordinary water was used. If this was correct, then the heating effect had nothing to do with nuclear fusion at all. It was simply a chemical reaction, the direct result of the electrical energy that had already been pumped into the cell over many hours. "If it is a chemical reaction, there's no way on God's earth we can be getting more out than we put in. It's like changing base metals into gold. It can't be done," said one of the scientists involved.

As Tom Bolton, director of Canada's tokamak fusion program put it, "The last thing the world wants is another source of lukewarm water!"

David Williams, who was conducting tests on cold fusion at the British nuclear laboratories at Harwell, was disappointed that researchers had been rushing to the press without even bothering to run a proper check on their results. "These people are engaging their mouths before they engage their brains," he said. In his opinion, the really serious research groups would not be saying anything until their experiments had been double-checked, and that meant waiting until the summer. Later Williams was to announce that his team had done "painstaking checks" but had detected no heat or radiation.

Stanley Pons's reaction to claims that cold fusion would not work was "Do your chemistry first." If the experiment is being done correctly, it takes two weeks to get enough deuterium into the palladium rod to produce fusion.

By April 21 the French-Swiss group, involving scientists at the French Nuclear Authority and the Lausanne Polytechnic School, announced that no neutrons had been detected in its experiment, but that the group planned to repeat them 2,000 yards under the Alps. In this underground environment, the background effects of most cosmic radiation would be filtered out so that it would be easier to detect any neutrons coming from the cell.

Fusion fever was not without its lighthearted side. On April 1 a British newspaper announced the successful harnessing of cold fusion by a group of students in their school laboratory. A television news team vigorously pursued the story until the dateline on the newspaper attracted their attention. The whole story was an April Fool's hoax.

Humor also came from a group of scientists at the Free University of Berlin and the German Chemistry Association's Frankfurt Institute, who called a press conference to announce that a patent had already been taken out on the process—all of sixty years ago!

Two Berlin chemists, Günther Marx and Waldfried Plieth, had been running an electrolysis experiment when their palladium electrode became so hot that it burned a hole in the wooden benchtop. With the assistance of a colleague, Gerhard Kreysa from Frankfurt, they needed only a short time to come with an explanation—and this did not involve nuclear fusion.

Oxygen and hydrogen react violently together. Indeed, this mixture is the basis of modern rocket fuels. The German scientists' idea was that oxygen from the atmosphere was entering the electrode, where it reacted with the absorbed hydrogen and generated great heat. In fact, this effect had been discovered in 1823 by Johann Wolfgang Döbereiner, who used it to power a cigarette lighter. The same process was later patented in the 1920s by two researchers at the University of Hamburg.

Explanations that the heat from cold fusion must have a simple chemical origin had already been flashing across the computer conference networks. But naturally this sort of effect would be known to any chemist who had worked with palladium, and the Utah researchers replied that it was certainly not an explanation for what they had been seeing for the last few years. "There is no conceivable known chemical reaction, even if you consumed all the matter in the cell by oxidation or hydrogenation, that could produce that much energy," Pons said. He pointed out that 10 million times more heat is released in nuclear fusion than in the chemical burning of the same amount of deuterium. A chemical reaction, Pons argued, could not account for the large amount of heat he was observing—it had to be fusion.

The first weeks of fusion fever must have been an exciting time for the Utah group, with Pons claiming that thirty groups had confirmed his result.

But now rumors of negative results and critical reactions were coming in. While many scientists were willing to agree that Steven Jones and the Brigham Young group had written a balanced scientific paper, the same could not be said of Fleischmann and Pons. A number of scientists were angered by what they felt were extravagant claims that had not been backed up with carefully recorded data, tables, graphs, or detailed accounts of experimental procedure. As a physicist at the University of Toronto put it, "I wouldn't accept a paper like that from a graduate student. Even the axes of the graphs are not properly labeled."

After several groups had tried in vain to detect any evidence of cold fusion, they began to realize how little everyone had been told about the actual process. Nathan Lewis of the California Institute of Technology claimed that he had tried in vain to extract information from Pons and Fleischmann. He had no idea of the exact conformation of the Utah cell and had to work from a glossy $8'' \times 10''$ photo. At one point, he attempted to estimate the dimensions of the apparatus based upon his guess as to the length of Stanley Pons's hand in the photograph! Later he was to learn that what he had been using was simply a publicity photograph and not the record of an actual fusion cell.

In the meantime, Brigham Young and the University of Utah were getting into some heavy legal disputes over patents. There was already bad feeling between the researchers at the two universities, and it looked as if things would be clarified only after the scientists and universities had gone to court. If anyone was going to make money out of cold fusion, it would be, in the short run, the lawyers. As Jim Brophy, vice president of research at the University of Utah, put it, "If there is practical application of this technology, then there is going to be either the world's largest patent lawsuit or the world's largest negotiation of settlement between people, each of whom thinks they [*sic*] have something."

Originally the University of Utah had hired a California attorney, Peter Dallinger, to file the first application, but now the Utah attorney general's office announced that the Salt Lake City firm of Giauque, Williams, Willcox and Bendinger had been hired to oversee the patent fight. In turn, Richard Giauque, "one of the finest trial lawyers in America," according to a press release on the computer network, had contracted with the Houston firm of Arnold, White and Durkee for assistance in assuring national and international rights.

James Brophy said, "Events which probably ought not to be made public have occurred in the last few days which indicate that there are people who are going to take every advantage they can." Richard Giauque said that his immediate goal was to stop the flow of information that could undermine the patents: "We need to get a tight hold very early on disclosures."

Already MIT was said to have filed its own patent applications on interpretations of the experiment. In fact, Giauque said, "As we scouted the major patent firms, we found that some of them had developed conflict-of-interest problems already."

QUESTIONS ARISE

Again and again physicists and chemists reported that they had attempted to learn more about the Fleischmann and Pons fusion process but had failed to obtain any response from the two researchers. Disillusionment over the whole subject of cold fusion was growing, and it was being directed not so much toward Jones and the

Brigham Young group as at Pons and Fleischmann. This feeling was to become focused in Baltimore, where on May 1 the first of two cold-fusion sessions was held by the American Physical Society.

The setting was the annual spring meeting of the American Physical Society, in the bulletin of which was buried, innocently enough, Steven Jones's talk entitled "Cold Fusion: Recent Results and Open Questions." Special press rooms had to be established at the Baltimore Convention Center. Everywhere that Steven Jones walked, he was followed by reporters, television cameras, and eager scientists trying to slip him copies of their papers or details of their own efforts on cold fusion.

At 7:30 that evening, almost 2,000 physicists and reporters met in a large assembly room to listen to a series of talks on cold fusion. Both groups, from Brigham Young and from the University of Utah, had been expected to attend, but in the end neither Pons nor Fleischmann showed up. The meeting itself was to last until after midnight, and because of the number of talks involved, it spilled over into the following evening.

After opening addresses by the chairman of the meeting and by the president of the American Physical Society, the audience settled back to hear the star of the evening, Steven Jones. Jones began modestly, arguing that he had achieved a "scientific breakthrough" with muon-catalyzed fusion as far back as 1983 while at the Idaho National Engineering Lab. At the time, the laboratory had urged him to issue a press release, but he had refused until the results were published. By then, as he pointed out, interest had fizzled. If only the same thing had happened with cold fusion!

Jones also gave the background to that curious mixture of salts that he had used in his cold-fusion cells, salts "characteristic of hot springs." After Jones had persisted with the idea of high-pressure cold fusion, a friend had given him the clue of trying to duplicate conditions within the earth—possibly geological heat was actually being generated through cold fusion. Steven Jones then showed the meeting a copy of a page from a laboratory notebook dated May 27, 1986, in which the idea of using electrolysis had first surfaced certainly long before he had heard of Pons and Fleischmann.

Jones reported on the successful detection of cold fusion at

Brigham Young University and answered questions put to him by the scientists in the audience. He did emphasize, however, that the Brigham Young results were showing that the effect was not large and that there would be no shortcut to fusion, no royal road to abundant energy. Yet at least his work had shown that a number of new possibilities had opened up for the harnessing of cold fusion. Clearly what Jones had seen was a very different phenomenon from that claimed by Fleischmann and Pons.

From that point on, the meeting was divided into a series of theoretical and experimental talks. The theoreticians did not seem at all fazed by the idea of cold fusion; in fact, they were willing to speculate on a variety of mechanisms that would allow deuterium nuclei to fuse and even to produce heat with little radiation. Listening to their talks, it was hard to believe that cold fusion was not happening all around us! Those with titanium implants in their bones or titanium implants in their teeth must have felt relieved that they did not regularly take heavy water with their whiskey.

The experimentalists were not, however, so liberal-minded, and talk after talk detailed their failure to find nuclear fusion, at least at the rate Fleischmann and Pons had claimed. It was only on the second day, with two reports from Argentina—one observing fusion, the other not—that the possibility of a large-scale fusion reaction was not ruled out. But by far the majority of scientific groups reporting at that meeting were coming down heavily against Utah fusion.

Over the two days, a number of experiments were reported in which great care was taken to rule out the effects of background radiation on sensitive detectors. Again and again the meeting heard the verdict "no neutrons," "no characteristic gamma rays," "no excess heat." It looked as if the meeting was out to bury the idea of cold fusion as an endless source of power.

Fleischmann and Pons had given three main characteristics for the fusion they had observed: excess heat, neutrons emitted from the fusion reaction, and gamma rays (being produced indirectly as the neutrons entered the surrounding water bath). These came from the following reaction:

Deuterium + Deuterium = Helium 3 + Neutron + Energy

There was also some evidence of tritium being produced as a by-product of the fusion reaction:

Deuterium + Deuterium = Tritium + Hydrogen + Energy

Although Pons and Fleischmann had emphasized the incredible amount of heat being produced, the speakers in Baltimore directed the focus of their attack against the other two signatures of cold fusion: neutrons and gamma rays.

Of course, Pons and Fleischmann had proposed that most of the fusion heat was being produced without any nuclear radiation. So the inability to detect neutrons or gamma rays did not totally demolish their claim. It would, however, have gone against Jones's theories. On the other hand, all these negative results did combine to shake faith in the reliability of Fleischmann and Pons's experimental procedures.

Nuclear Fusion or Basement Radon?

The gamma rays reported by Fleischmann and Pons were easiest to dispose of. A number of physicists had been suspicious about them from the beginning. The spectrum reported by Fleischmann and Pons did not look quite right, the energy did not appear to be in the right place and the plate was too narrow—a fact that was crucial to claiming the rays resulted from deuterium-deuterium fusion.

A tricky piece of scientific detective work tracked the gamma rays down to contamination in the Utah basement laboratory. Basements are notorious places for accumulating radon gas—a naturally occurring radioactive substance that filters through cracks in the bedrock. During its normal radioactive decay, radon produces another isotope, radioactive bismuth 214, and this bismuth emits a gamma ray with exactly the same characteristics as that seen by Pons and Fleischmann. In fact, scientists said, the gamma rays they were detecting had nothing to do with nuclear fusion and were simply natural background radiation in the laboratory. They pointed out that such radiation could vary strongly from location to location and would be affected by the air conditioning, which would sweep the radon gas out of the room.

Neutrons or Not?

At the APS meeting, group after group emphatically claimed that it had not seen an increase in the number of neutrons. Neutrons are always present, coming from natural background radiation and from the cosmic rays that constantly bombard the earth. But the number of neutrons in this natural background fluctuates from day to day. The essential question was, had Fleischmann and Pons really seen additional neutrons, or was what they saw just the result of some chance effect? The physicists argued for chance.

Not everyone denied seeing neutrons. The group from Yale University, for example, reported seeing neutrons—but at a rate one million times smaller than that reported by Pons and Fleischmann. Indeed, during one run, the Yale researchers observed only two neutrons—which the group then named Pons and Fleischmann. While a number of research groups could not rule out a very low level of nuclear fusion—possibly at the level that Steven Jones claimed to have witnessed—it was certainly nothing like that claimed at the University of Utah.

Heat or Illusion?

Pons and Fleischmann had pinned the flag of their discovery on the excess heat observed in the fusion cell. This heat was so great that it could never be explained in chemical terms alone. Even if gamma rays turned out to be a natural effect of the laboratory itself and few neutrons were observed, the whole point of the cold-fusion effect was the enormous energy produced. Was the significance of this finding going to evaporate as well?

The main attack on Pons and Fleischmann came from Nathan Lewis of Cal Tech and W. E. Meyerhoff of Stanford. As it turned out, this excess energy was far from straightforward to observe. Researchers could not simply measure the amount of heat being given out, since a great deal of energy was constantly being pumped into the system in the form of the electrical current that drives the cell.

The key to Fleischmann and Pons's claim was that more heat was coming out than was going in. The situation is rather like a busy subway station with passengers bumping into each other and pouring in and out of the entrance. In all that flurry of movement, it could be difficult to discover whether more people were leaving than entering the subway.

The act of discovering how much energy is being created depends on how you do the calculation. Lewis claimed that by slightly changing the assumptions made about the cell, he could get a result that indicated that energy was actually being taken in! The fusion cell then, he claimed, was no power source; it was really acting like a refrigerator, absorbing energy rather than creating it.

W. E. Meyerhoff argued that there is a difference in the temperature of the two electrodes in the cell. While the temperature is raised at one palladium electrode, it is lower at the other. This effect has nothing to do with nuclear fusion and is simply a natural effect of electrolysis, the result of the electrical current breaking up the water molecules.

If the heavy water in the cells is not constantly stirred to mix up the hot with the cold water, then—even without any cold fusion—a high temperature would be registered close to one electrode, with a lower temperature close to the other. In fact, depending on where the thermometer was placed, you would get totally different answers about the amount of heat produced. It would be like standing between a hot stove and a refrigerator—move a little bit to one side or the other, and the temperature goes up or down.

Meyerhoff pointed out that Fleischmann and Pons had not stirred the heavy water in their cell to even out the temperature. The thermometer had been placed close to the palladium electrode where it naturally registered a high temperature. His conclusion was that Fleischmann and Pons had erroneously assumed that this same temperature was occurring throughout the cell and had incorrectly calculated that a great amount of heat was being given out. If the thermometer were shifted toward the other electrode, Meyerhof argued, the temperature would appear to drop and indicate that heat was not being produced.

In other words, Fleischmann and Pons did not really know how to do a good experiment or how to calculate their results. This was a damning accusation indeed.

Figure 5-1   THE APS ATTACK

Neutrons   Careful experiments carried out by a number of
                  groups detect either no excess neutrons or levels
                  millions of times lower than Fleischmann and Pons
                  had claimed.

Gamma rays   This radiation is not being produced as a result of
                  cold fusion. It is simply the by-product of radon gas
                  in the laboratory.

Excess heat   No excess heat is detected by a number of groups. It
                  is claimed that the energy emission heat detected by
                  Fleischmann and Pons is simply the result of the way
                  they did their calculation. Another claim made is that
                  Fleischmann and Pons did not stir the heavy water in
                  their cells, so that different temperatures built up at
                  different points in the cell. Pons counters this
                  criticism by showing how rapidly a drop of dye mixes
                  into the system.

A week later, at a meeting of the Electrochemical Society, Pons was to refute this most serious of attacks on their experiment—that heat registered in their cell was simply a result of poor mixing of the heavy water. In a short movie, he showed a drop of red dye injected into the fusion cell. Within twenty seconds, the dye had swirled through the water, indicating that the solution was being thoroughly stirred by the reactions occurring inside. In such a cell any differences in temperature would be averaged out, and Meyerhof's main objection appeared to have been countered. Lewis, however, remained unconvinced.

But scientists at the American Physical Society meeting did not know about this rapid mixing effect, and after they had heard how many groups had failed to confirm cold fusion, the general feeling was that Pons and Fleischmann were mistaken. The strongest statement came from Steven Koonin, a theoretical physicist from the University of California at Santa Barbara: "We are suffering from the incompetence and perhaps delusion of Doctors Pons and Fleischmann."

At a press conference held on the second day, Steven Jones asked a panel of scientists to vote on the Utah result. The verdict was eight to one against. The dissenting voice came from John Rafelski, one of

Jones's co-workers, who argued that it was too early to be definite about these experiments and that it was still possible that something more interesting was happening.

In the opinion of Douglas Morrison, a scientist from CERN who had followed the cold-fusion story from its beginning and was editing an unofficial "Cold Fusion News" on the Bitnet computer network, the subject of cold fusion was now more or less closed. The verdict had gone strongly against Pons and Fleischmann—no excessive heat or nuclear radiation was being produced in their test tubes. As to Jones's result, that seemed to be an open question. Possibly cold fusion was a reality, but the reactions themselves were occurring on the margin and could not yet be used as a power source.

IN THE AFTERMATH OF APS

Not everyone was happy with the way the APS meeting had gone. Uziel Landau of Case Western Reserve said, "I think the statements were just outrageous." Landau respected Fleischmann: "He's a very cautious, very careful guy, highly regarded. I take very seriously any statement that has come from him.

Researchers at Texas A&M were also "a little surprised" at the acrimony of the meeting. By this time, a second group, under John Appelby, director of the Electrochemical Systems and Hydrogen Research Center, had found excess heat.

Likewise, scientists at Stanford University were persisting in their claim to have detected excess heat, but they anticipated criticism. Said a Stanford spokesperson, "They are sticking by their guns . . . (but) they are circling the wagons."

Walter L. Peterson, Jr., of San Diego, speaking over the computer networks, listed the following headlines:

- "Professor claims radical new discovery—scientific community skeptical"

- "Nation's top experts disagree on 'discovery'"

- "Chief of nation's leading lab can't reproduce experiments"

- "Prof repeats claim of new discovery—will disclose more details of experiments"

- "Scientists still not convinced—say discovery is only speculation"

- "Prof offers to help reproduce experiments"

While these headlines may seem to recount the story of Fleischmann and Pons, they actually are imaginary reports of the true story of Sir Isaac Newton and his "crucial experiment" on the composition of light. If Newton had been living and working in our era, the newspapers might have issued such reports concerning his new idea.

On February 19, 1671, "A Letter of Mr. Isaac Newton containing his New Theory of Light and Colours" appeared in the *Philosophical Transactions*. In this publication Newton claimed that he had split sunlight into its component colors using a prism. But Robert Hook, curator of experiments for the Royal Society, could not reproduce Newton's finding. Indeed, other members of the society began to doubt whether Newton had ever performed the experiment successfully. There was even a rumor that the whole thing was a guess and had never been carried out. "Newton fever," if one may turn a phrase, spilled over into Europe, where a number of leading scientists, including Huygens, cast doubt upon the whole idea.

The controversy dragged on for years, with some scientists still doubting that light could be composed of colors. Today, of course, Newton's experiment is taken for granted, although it must be pointed out that few people outside a teaching laboratory ever take the trouble to repeat it.

The analogy that Walter Peterson was making between Newton and Fleischmann and Pons is clear. In both cases, a relatively simple experiment demonstrates an absurd and counterintuitive result. The conclusion of the experiment shakes scientists' faith in what they had assumed about nature, and attempts to reproduce the result arouse controversy. Possibly that experiment was not so simple after all, or could the whole thing be a hoax, a piece of scientific fraud?

With the close of the Baltimore meeting, the heat had gone out of the cold-fusion story. Newspapers continued to carry updates, but these came generally from the wire services and did not merit the attention of a special reporter sent to follow the principal researchers across the United States. Although the *New York Times* could not resist a parting shot, saying that the University of Utah "may now claim credit for the artificial heart horror show and the cold-fusion circus, two milestones at least in the history of entertainment if not of science." (The reference to the artificial heart refers to the world's first implantation of an artificial heart, which was performed in Salt Lake City.)

Even Pons and Fleischmann appeared to be out of favor in Washington. On one occasion they flew to Washington to meet John Sununu, White House Chief of Staff, who then canceled the meeting. But despite the negative feeling at the APS meeting, it appeared that James D. Watkins, U.S. Secretary of Energy, had directed his laboratories to step up their investigations of cold fusion. A special panel was to be established to investigate cold fusion. It would consist of experts in electrochemistry, nuclear physics, solid-state physics, and engineering.

THE CHEMISTS' TURN

While the physicists had attempted to demolish Fleischmann and Pons at Baltimore, it was now the turn of the chemists to give their version of the story at Los Angeles, where on the following week, the Electrochemical Society was holding its annual meeting. There at least, Pons and Fleischmann would have the chance to defend themselves.

According to Hugo Rossi, who had been appointed interim director of the University of Utah's Cold Fusion Research Project, May 8 would be "F-day," the day on which the chemists would take their revenge on the physicists. With new confirmations coming in, Rossi hoped that the state's Fusion Energy Advisory Board would free the $5 million for research.

In fact, the San Diego meeting appears to have been fairly low-key. Although Pons and Fleischmann were not about to back down, Martin Fleischmann was willing to concede, "I have always been ready to acknowledge the fact our experiments could be faulty. If we turn out to be wrong, I'll be the first to admit it."

But Stanley Pons was in a fighting mood: "We are extremely pleased because they confirm our findings. The absence of neutrons doesn't concern us in the slightest. We couldn't be happier. We and other scientists will soon tell them why this is so."

By this time Nathan Lewis of Cal Tech had turned out to be the star critic of cold fusion. "There is no fusion," he emphasized. Again Pons countered, "We are amazed that Professor Lewis has learned how to solve all these problems in only one month, when it took us five and a half years."

Ralph Brodd, former president of the Electrochemical Society, cautioned, "I don't think anything has been proved. It's going to take time to sort it out."

But researchers at Stanford University had checked their own measurements of heat output and stood by their earlier confirmation. Texas A&M reported a neutron measurement that was significantly above the background and 80 percent more heat coming out of the cell than energy put in. If the physicists in Baltimore, where it rained heavily over the two days of the cold-fusion meeting, had been almost unanimous in their rejection of Pons and Fleischmann, the mood was a little different in sunny San Diego, where electrochemists gave a generally favorable reception.

For their part, Pons and Fleischmann were announcing ten times as much heat production as a few weeks earlier and the phenomenon of seeing "sudden bursts of heat" fifty times above the break-even level. Tritium levels were three times higher than background. Pons also announced that a collaborative experiment was being arranged with Los Alamos, and the results of other test experiments would be announced in a matter of days.

The Utah group had begun to charge up additional cells, which were due to be transported to Los Alamos for testing. However, a three-hour power failure in the laboratory set their work back by three to five days, and they had to begin charging again.

Rumors were also circulating that Fleischmann had not told all and that a number of interesting new developments were yet to be announced. It was said that Fleischmann was smiling up his sleeve, looking at the negative results that were coming in and simply saying nothing. Those failures to observe great quantities of cold fusion arose

BROKEN VISIONS
Broadcast Power

At the turn of the century, Nikola Tesla claimed that he could broadcast electrical power over great distances without loss and without the need for wires. Tesla's credentials were impeccable, for he had invented the AC electrical motor and had harnessed the power of Niagara Falls to produce electrical energy.

"Tesla transmitters" were built in Colorado and on Long Island, New York. Tesla was even supposed to have broadcast power without wires between Canada and the United States. Claims to have achieved Tesla transmission still surface from time to time but are discredited by scientists.

N-Rays

In 1903 the French scientist R. Blondlot announced that he had discovered a new form of radiation, N-rays, being given off by metals and even by humans. The curious x-rays had recently been discovered, and N-rays seemed to have even more bizarre properties—they appeared to decrease with loud noises, and some scientists claimed they were being absorbed from the sun.

Eventually the idea of N-rays was thoroughly debunked, but support for the phenomenon continued in some quarters.

Hot Fusion

In 1958 British scientists announced that they had successfully harnessed nuclear fusion in a reactor called ZETA. Neutrons were detected coming from the hot plasma, and the noted scientist Sir John Cockcroft announced that he was 90 percent certain that fusion had occurred. "Limitless Fuel for Millions of Years," hailed the London *Daily Mail* in its headlines.

Later it was discovered that these neutrons were being produced by another process, and hot fusion would lie many decades away.

Polywater

One of the most exceptional of scientific delusions was polywater, a supposed new form of water with a higher freezing and boiling point. Polywater was first "discovered" in the Soviet Union, and from 1968 to 1972, its properties were confirmed by laboratories all over the world, including the prestigious National Bureau of Standards.

However, the existence of a new form of water was finally put to rest when it was discovered that polywater was simply plain old water with a few impurities.

Water Memory

On June 20, 1988, Jacques Benveniste, a French scientist, and his colleagues published a paper in *Nature* claiming that water had a "memory" for biological activity. An active substance was added to water and then diluted so many times that the chance of finding one of the original molecules was one in 10^{96} (a 1 followed by ninety-six zeros). But, despite the absence of any of the original substance, the water retained its biological activity. Benveniste's result, which appeared to provide an independent confirmation of homeopathy, was confirmed by laboratories at the Kaplan Hospital in Rehovot, Israel, and at the University of Toronto. (Homeopathy is a form of medicine in which active substances are administered in vanishingly small dilutions.)

Although it published the article, *Nature* was not satisfied and sent a team including the magician Randi and Walter Stewart (who had investigated a number of alleged scientific frauds) to Benveniste's laboratory. On July 28 *Nature* reported their results in an article entitled "High Dilution Experiments a Delusion." Benveniste and his supporters, however, still maintain that the effect is authentic.

because the scientists involved did not know how to run the fusion cell properly. Apparently Fleischmann and Pons were scheduled to produce a comprehensive report in the summer.

In addition, they had entered into an agreement with Johnson Matthey—worldwide supplier of precious metals in scientific grades. The palladium electrodes would be returned to the company for analysis. It was hoped that this would settle some serious arguments about the presence of helium 3 and tritium—fusion by-products that would confirm their claim.

A COLD-FUSION WORKSHOP

Although Fleischmann and Pons were conspicuously absent, the Workshop on Cold Fusion Phenomena, held at Los Alamos on May 23–25 and broadcast across North America by satellite, proved to be one of the most balanced and informative discussions of the whole subject. At last scientists from all over the world had the time to discuss the details of particular experiments—with the noted exception of those carried out at the University of Utah.

This meeting was sponsored by the U.S. Department of Energy and the Los Alamos National Laboratories and allowed not only the more than 400 scientists registered at the meeting, but also the thousands of others who watched the proceedings via satellite, to explore the whole issue of cold fusion.

A variety of results emerged from the Los Alamos Conference, which appeared to sum up the various positions on cold fusion.

Pons and Fleischmann Fusion

Enormous quantities of energy are produced in the palladium electrode after running the cell for times exceeding 100 hours. This, it is claimed, can only be the result of a nuclear reaction. Excess heat is the characteristic of Pons and Fleischmann fusion, but the neutron counts involved are millions of times lower than expected on the basis of a conventional nuclear reaction and suggest some new, exotic fusion reaction.

Figure 5-2

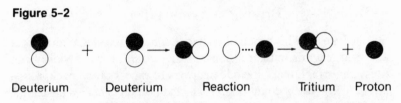

Deuterium   Deuterium   Reaction   Tritium   Proton

The protons in the two deuterium nuclei are kept apart by their repulsion. The favored reaction will lead to the production of tritium and a proton and *not* helium 3 and a neutron. Such a process would explain the low rate of detection of neutrons during the fusion process.

Supporters of Fleischmann and Pons fusion argue that those experimenters who do not observe this heat are not doing the experiment properly. The correct experiment has the following requirements:

- Pumping deuterium into the electrode for a long enough time

- Using large electrical currents and the correct voltages

- Using the right form of palladium and preparing the electrode correctly

- Putting the correct additives in the cell solution in order to "poison" the electrode

- Luck

Support for Pons and Fleischmann fusion came from a number of countries. The strongest scientific support in the United States came from Texas A&M.

Frascati Fusion

Deuterium gas is driven at high pressure into a metal—titanium or palladium. Bursts of nuclear activity are observed as the metal warms up from liquid-nitrogen temperature. Variants of this approach involve heating the metal to high temperatures or subjecting it to the focused force of a chemical explosion.

Brigham Young Fusion

Steven Jones and his group observe a number of the characteristics of a nuclear reaction in their fusion cells after running for only one hour. Most important, these include neutrons in excess of the background levels. The neutron count is not high, and no excess heat is measured.

Various aspects of Brigham Young fusion have been confirmed by a number of groups, but the phenomenon is vastly smaller in scale than that claimed by Pons and Fleischmann and has no direct application as a commercial source of energy. Supporters of Brigham Young fusion tend not to believe in Pons and Fleischmann fusion.

No Fusion

A number of groups, after having carried out extremely careful experiments, see no evidence of nuclear fusion. No excess heat is measured, no neutrons are confirmed, and none of the products of nuclear fusion are detected. If any effect is observed in a cold-fusion cell, it is probably the result of some nonnuclear process and certainly is not a new source of energy.

Critics of this group argue that their experiments have not been carried out properly. For example, currents are too low, times are too short, electrodes are not properly prepared, and incorrect solutions are used in the cell.

One of the strongest speakers at the Los Alamos conference was John Appleby from Texas A&M, who not only confirmed the Pons and Fleischmann phenomenon but also presented experimental results to counter a number of objections—for example, that the heat was being produced by chemical recombination of oxygen and deuterium. Appleby, however, did not detect any helium in his heat-producing cells.

The results of his experiments are shown here in a series of graphs. The first shows that if platinum is used in place of a palladium electrode, or ordinary water in place of heavy water, no heat is seen. However, as shown in the second graph, a definite rise in excess heat is observed with deuterium pumped into palladium.

One of the magic recipes in a Fleischmann and Pons cell is the

Figure 5-3

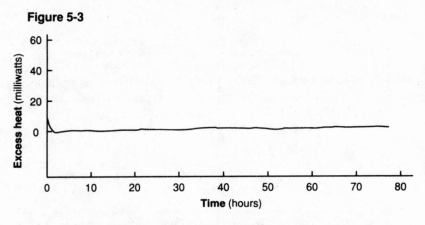

Representation of John Appleby's results at Texas A&M. Cold fusion cell with a platinum electrode in place of a palladium electrode. No excess heat is observed.

Figure 5-4

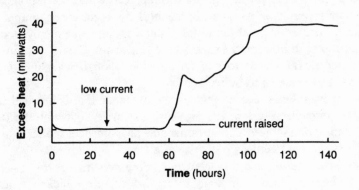

The Texas A&M results with a palladium electrode in heavy water. Heat is evolved.

Figure 5-5

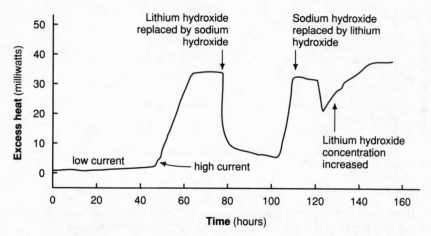

John Appleby observed that when Lithium hydroxide is removed from the cell and replaced by sodium hydroxide, the heat falls. What role is the lithium playing in the cell?

addition of lithium in the form of LiOD. In a particularly striking Texas A&M experiment, lithium was replaced with a corresponding sodium preparation. As shown in the third graph, the energy output of the cell immediately began to fall. When sodium was removed and replaced with lithium, the excess heat rose again. Clearly lithium is playing an important role in cold fusion, though what could be happening remains to be worked out.

Steven Jones also presented new evidence that confirmed his own approach to fusion. His experiments carried out with the Los Alamos group showed an interesting new phenomenon: sudden bursts of neutrons—sometimes over fifty at a time.

The meeting heard of neutrons being confirmed from fusion cells that were located in a laboratory 1,400 meters under a mountain in Italy, an experiment conducted in conjunction with BYU. At such a depth, cosmic rays and background neutrons have very little effect. Experiments like these also rule out the possibility that cold fusion has anything to do with muons. At such depths there simply are not enough muons around to begin a fusion reaction.

There were also reports of careful experiments in which absolutely no effect was seen. On a number of occasions, John Bockris of Texas A&M stood up and argued with invited speakers that the negative results simply resulted from doing the experiment incorrectly. The researcher should first generate the phenomenon of excess heat and then worry about detecting neutrons, tritium, and helium.

In the end, one member of the audience had had enough and complained about always being castigated for not doing correct experiments. If there was a right way, then why weren't Pons and Fleischmann telling everyone—why were they keeping their work secret? "I'm getting tired of seeing science conducted by press release," he said.

By the end of the Los Alamos workshop, opinions on cold fusion were divided, but at least everyone agreed that pumping deuterium into palladium or titanium was producing a variety of effects that no one really understood. Even the strongest cold-fusion critics agreed that, as scientists began to focus on deuterium in metals, they would be entering a new and exciting field.

Those who supported cold fusion believed that they had even stronger evidence than before, although opinion was strongly divided on whether the Fleischmann and Pons experiment had really generated high excess energy. Many researchers voted for the low levels of nuclear fusion seen at Brigham Young University.

Irrespective of the Los Alamos conference, the University of Utah was pressing ahead with the idea of developing commercial energy from cold fusion. Pons and Fleischmann had set up a new series of experiments and were planning a major publication for the summer. In addition, a group led by Dr. Milton E. Wadsworth of the College of Mines and Earth Sciences, and working quite independently of Pons and Fleischmann, had confirmed the heating effect.

Wadsworth's idea was to carry out a thorough metallurgical examination of the palladium electrodes to discover what was happening inside. For this he would need some fusion cells of his own. After working for almost a month, he was ready to give up, when his apparatus recorded some sudden excursions of heat. These increases in heat were recorded again and again and lasted for over an hour. Clearly something dramatic was going on within the electrode.

Wadsworth and his group were also in a position to analyze the surface composition of the electrodes that had failed. He discovered that their surfaces were contaminated with platinum (from the positive electrode), copper (from the electrical connections), and silicates from the glass vessel. It evidently was critical to prevent the poisoning of the surface of the electrode, and this meant running the cell under exactly the right conditions.

In later experiments he would study the interior of the metal to discover if microcracks were forming. Other groups planned to search for the products of a nuclear reaction such as helium 4, helium 3, and tritium.

But was a high level of cold fusion really going on inside Wadsworth's electrodes, or could all that heat still be explained by some other process? And when all those emissions of heat were added up, would they really outweigh all the energy that had been put into the system during the charging process? In other words, could a Pons and Fleischmann cell be made to run itself and still produce energy? That was the question the scientists were asking.

As far as Fleischmann and Pons were concerned, they had already confirmed a net heat output. For the two Utah scientists there were no more questions; the age of cold fusion had truly arrived.

Chapter 6
Cold Fusion: From Sun to Earth

Cold fusion is a hot topic that is going to be debated for a long time to come. On the one hand, it is a really exciting new scientific discovery that forces chemists, physicists, engineers, and materials scientists to work closely together. Cold fusion brings together so many different skills and disciplines that the only promise of success lies in a truly integrated approach.

But beyond the purely scientific interest in cold fusion is the possibility that this phenomenon, if it does indeed exist, may one day be developed to such a point that energy can be extracted on a large scale. If this happens, then cold fusion will certainly have been one of the most exciting scientific discoveries in decades, possibly in centuries—as one scientist put it, "the most important discovery since fire."

With the Utah press conference of March 23, 1989, the idea of cold fusion had burst upon the world. Within a matter of hours, people were hearing about a new and dramatic form of energy, a virtually inexhaustible supply that would be cheap and pollution-free. By the beginning of June 1989, hardly two months later, many scientists were willing to concede that something very significant was happening. While a number of groups had failed to confirm the phenomenon, other laboratories all over the world had definite evidence of at least some form of cold fusion.

Steven Jones was claiming that cold fusion had indeed been established but only at a very low level. As far as he was concerned, the heat produced by Pons and Fleischmann had nothing to do with nuclear fusion and would therefore never represent a source of inexhaustible power. But Pons and Fleischmann were convinced: cold fusion was here to stay and would eventually be scaled up to full commercial production.

As details began to surface about the University of Utah and Brigham Young University experiments, some newspaper editorials began to question why conventional science had been pursuing the wrong track in its attempt to harness high-temperature fusion. For decades, large sums of money had been poured into tokamaks and inertial confinement systems, but now a totally new idea had come along.

But cold fusion had been around since the experiments of Luis Alvarez, and there were hints that scientists had been thinking about exotic nuclear reactions half a century ago.

HELIUM CREATION

On September 17, 1926, Reuters news agency carried a report that two German scientists had, after years of experimentation, succeeded in transforming hydrogen into helium "with the aid of particles of metal." Two respected German scientists, Fritz Paneth and Kurt Peters, working at the Chemical Institute of the University of Berlin, believed they had evidence for nuclear fusion within palladium.

Of course, the whole idea of nuclear fusion was not properly understood at that time; quantum mechanics was only a year old, the neutron had yet to be discovered, and the fusion of hydrogen nuclei as the source of energy in the stars had not yet been worked out. Scientists did know that a curious new element, helium, could be found in the sun's atmosphere but was exceedingly rare on earth. This helium had to be a by-product of the energy-producing reactions that occur inside the sun itself.

What Paneth and Peters had done was to pass hydrogen gas through a thin, red-hot tube of palladium. In the gas that came out the other end, Paneth and Peters detected a very small amount of helium. Somehow helium was actually being created out of hydrogen inside the palladium—curious anticipation of the Frascati fusion of 1989! The two scientists persisted with their experiments and discovered that they could increase the amount of helium by using other preparations of palladium.

The next step was to see if there was any possibility of a mistake. After carefully searching for all possible sources of error, Paneth and Peters published their results in 1926. Helium was being created out of hydrogen by some unknown fusion process.

But in April of the following year, a retraction appeared. The two scientists had discovered that tiny amounts of naturally occurring helium were absorbed on the surface of the glass vessels used in their experiments. When this glass was heated, the small amounts of helium were released. They also discovered that one of the catalysts they had used in the experiments—platinized asbestos—gave off absorbed helium when heated. In other words, they could not be 100 percent certain that the helium they had detected was actually being "created" by nuclear fusion, rather than being simply the result of contamination.

In February 1927, however, a Swedish scientist—John Tandberg of the Electrolux Research Laboratory—filed a patent for "a method to produce helium and useful energy." The process was based on the effect discovered by Paneth and Peters, but Tandberg claimed to have discovered a method to "significantly increase the efficiency in order to produce useful energy." This patent was not granted, but Tandberg and his colleague, Torsten Wilner, continued to experiment with heavy water—samples were obtained from Niels Bohr. The Swedish notebooks on cold fusion are now in the possession of Bertil Wilner, son of Tandberg's collaborator.

With the explosion of interest in nuclear and elementary-particle physics, and the new theoretical tools of quantum mechanics and quantum field theory, those rumors of cold fusion were simply forgotten. But hints that something curious could be happening inside metals still surfaced occasionally. For example, in 1978 three scientists from the A. F. Ioffe Physiotechnical Institute in Leningrad announced that they had discovered an unusual concentration of the rare isotope helium 3 in certain metals. Helium 3 is produced during one of the possible fusion reactions between deuterium atoms:

Deuterium + Deuterium = Helium 3 + Neutron

B. A. Mamyrin, L. V. Khabarin, and V. S. Yudenich knew that most of the helium in the universe occurs in the form of helium 4 (an atom containing two outer electrons and a nucleus of two protons and two neutrons). But there is also an alternative isotopic form of helium called helium 3 (also written ^{3}He), in which the central nucleus has two protons but only one neutron. Within the earth's crust, only around one in every 10 million (or even fewer) helium atoms occurs in this rare form; a similar proportion is found in the earth's atmosphere.

The three scientists discovered that helium gas that had become trapped in a number of pure metals contained a much higher proportion of helium 3. A number of different metals were heated to drive out the absorbed helium, and the amount of helium 3 and helium 4 was measured. To the scientists' surprise, many of the 300 different samples studied contained very large amounts of helium 3 and no detectable helium 4!

This was a staggering result, for it meant that the amount of helium 3 was up to a million times higher than normal. Moreover, the Soviet scientists discovered that this isotope was not uniformly distributed throughout the metal. It almost looked as if it had been created in tiny patches—around 1 cubic millimeter in size.

Where was this helium 3 coming from? In the view of Jones and his co-workers at Brigham Young University, the answer was staring everyone in the face. It was the direct result of nuclear fusion. A variety of gases, including deuterium, are naturally absorbed within metals. Suppose that a spontaneous nuclear fusion of this deuterium begins. The result of this fusion will be the creation of that rare isotope helium 3.

One possible fusion reaction is:

Deuterium + Deuterium = Helium 3 + Neutron

Another possibility is:

$$Deuterium + Deuterium = Tritium + Proton$$

This reaction generates tritium, the heavy isotope of hydrogen. Was there also evidence for unusual concentrations of tritium in metals?

Sure enough, this rare isotope of hydrogen was also found. Could nuclear fusion have been going on under our noses all this time without scientists noticing it, a perfectly natural process that occurs in many metals?

Steven Jones and the Brigham Young group had also talked to Harmon Craig at the University of San Diego, who gave them information about some other helium 3 anomalies. For example, Craig had discovered curious pockets of helium 3 in diamonds that had been cut by a laser beam. It appears that while absorbed helium 4 is distributed quite uniformly throughout the crystal, the rare helium 3 isotope occurs in tiny, localized pockets. Could these be associated with local areas of compression in which deuterium atoms fuse to produce helium 3?

These various clues are the circumstantial evidence that cold fusion may not be as bizarre or obscure as everyone had thought. Indeed, fusion may be a perfectly natural process in a variety of systems but one that had simply escaped widespread notice until March of 1989!

GEOLOGICAL HEAT

The Brigham Young group produced an even more impressive argument for cold fusion, and this came from the earth itself. Perhaps the geological implications of cold fusion will revolutionize our understanding of the planet.

The earth beneath our feet is never at rest; the foundations of the planet are far from firm. Every day, somewhere in the world, earthquakes shake the crust of our planet; hot springs pump their thermal energy above the ground and in some regions of the world are used for domestic heating; geysers of hot water spurt into the air at regular intervals; from time to time volcanoes become active, throwing out molten rock and firing hot gases high into the atmosphere. Anyone who has been down a coal mine knows that the deeper one descends, the higher the temperature rises until, in the deepest mines, it becomes uncomfortably hot.

Figure 6-1   THE EARTH'S HEAT

Hot springs: $20-100°$ C

Fumaroles*:   up to $560°$ C near Mt. Vesuvius

Lava: $700-1,200°$ C

Mines: A rise of $1°$ C for every 40–170 ft. deep; highest recorded temperature is $1,540°$ C in a 20,000-ft. well in Wyoming

Earth's core: estimates of $1,600-76,000°$ C; most probably $3,500-4,500°$ C

---

*Superheated steam emerging from vents in the earth

Divorced from our direct experience, but equally important, are the motions of the continents themselves. Since its creation, the earth's crust has been constantly on the move. The world's continents, as we now know them, split from a single protocontinent and for hundreds of millions of years floated across the globe on their tectonic plates. The whole upper crust of the earth is constantly being driven by the heat of the underlying mantle. Where tectonic plates meet, there is considerable geological activity, and given enough time, the earth's oldest rocks will be sucked down into the earth's mantle. Tens of millions of years later, melted and reformed, they will be thrown up into some new location.

The geologists' challenge is to discover the power source for all this activity. What drives the earth's mantle and the constant motion of the continents? What powers volcanoes and hot springs? Why does the temperature rise as one journeys deeper into a coal mine?

In the past, scientists tried to come up with a variety of solutions to this puzzle. Today the most striking solution has been proposed by Brigham Young University: our earth may, in part, be powered by cold fusion deep within its foundations.

In the nineteenth century, scientists believed that the earth had been formed as a molten body. While its outer crust had cooled over time, the interior was still molten and giving out heat. By measuring this flow of heat and working out the speed of cooling, it should be possible to calculate the age of the earth. But the great scientist Lord Kelvin discovered that this result did not make sense. The earth was too hot! Either our planet was far younger than anyone believed, or heat was still being generated inside.

Scientists therefore concluded that there must be a source of heat within the earth's mantle, something that was keeping the temperature high. But what could this be? Chemical processes were not the answer.

Only with the discoveries of nuclear physics did the picture begin to emerge. Scientists knew that a variety of radioactive elements occur naturally in the earth. As each of these disintegrate, it gives out a little heat. Would this natural radioactive decay be sufficient to heat the earth's mantle?

Nuclear energy in the earth is released in the decay of isotopes like uranium and its nuclear neighbor thorium. But these isotopes are only present in trace amounts. More abundant is radioactive potassium—potassium 40—and it is from this source that most of the earth's heat is supposed to come.

But the whole question of the earth's heat is highly complex, and any attempt to make a theory work involves many assumptions. In short, the question of what powers the earth and creates the various "hot spots" in its crust is still open. Some geologists do not believe that the radioactive decay of potassium can supply sufficient heat and that a "hidden source" of heat must be present. In fact, it was the search for an answer to this very question that first led Paul E. Palmer, a physicist at BYU, to suggest that Steven Jones build an electrolytic cold-fusion cell. The idea was to duplicate the chemistry of natural hot springs by adding a complicated "chemical cocktail" to an electrolytic cell of heavy water.

If Palmer is right and heat in the earth's mantle is being generated by cold fusion, then this will revolutionize geology and create a totally new picture of the interior of the earth. In fact, Paul Palmer of Brigham Young's physics department feels that this is one of the most exciting aspects of the cold-fusion work—the chance to solve one of the great puzzles of the earth. Already a number of clues to cold fusion lie beneath our feet.

HELIUM 3 UNDER THE OCEAN

If cold fusion is a reality, then the heating process would take place as follows: Where the tectonic plates meet, rocks, containing water, are drawn down into the mantle. There, under conditions of high temperature and pressure, nuclear fusion of the hydrogen isotopes in the water takes place. The chemical and physical structure of rocks and minerals inside the earth assists in the fusion process—as does the palladium electrode in a fusion cell. The result is a release of heat, along with the decay products of the fusion reactions—helium 3 and tritium.

The best evidence for cold fusion would be to look for super-heated plumes of water and determine if they are particularly rich in tritium and helium 3. In 1981 an important result came from deep in the Pacific Ocean at a latitude of 15° S on what is known as the East Pacific Rise. At certain points in the ocean floor, great thermal vents emit superheated water from the earth's mantle into the lower parts of the ocean. One of these vents is located some 2,500 meters beneath the Pacific Ocean in a region where the tectonic plates meet.

John E. Lupton and Harmon Craig, at that time at the Scripps Institute of Oceanography, were members of an expedition to this East Pacific Rise and collected bottles of water from the plume, comparing it with samples in the same neighborhood. Although the idea of cold fusion had not been contemplated at the time, the two scientists analyzed the water and discovered that it was abnormally high in helium 3. In fact, high levels of this by-product of nuclear fusion were still detectable for thousands of kilometers from the ridge.

Other concentrations of helium 3 have been found at vents near the Galapagos Islands and in the East Pacific at 21° N.

TRITIUM IN VOLCANOES

If nuclear fusion is occurring inside the earth, then not only helium 3 but also tritium must be produced. For reasons unconnected with cold fusion, a tritium-monitoring station was operated at the Hawaiian island of Mauna Loa during the 1970s. As the scientists had expected, a tiny quantity of naturally occurring tritium was measured, and its value fluctuated very slightly from day to day.

To the scientists' surprise, a sudden peak occurred during February–March of 1972. At the time, it was difficult to produce a convincing explanation for this effect. The only event that coincided with it was an eruption of the volcano on the island. Could it be that tritium was being shot into the atmosphere from the earth's interior?

Other tritium peaks were later discovered to coincide with volcanic activity on the island, but these data had to be discounted, since hydrogen bomb tests being carried out at the same time were also releasing tritium into the atmosphere. Currently the Brigham Young group is planning a new expedition to Mauna Loa to look for the products of cold fusion.

Figure 6–2

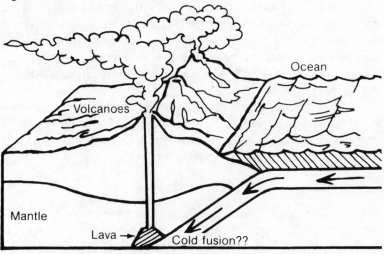

With helium 3 and tritium both being thrown out from the earth's mantle, could it be that cold fusion is the power that moves the earth? If this is true, then geologists are going to have to rewrite their textbooks and rethink everything they know about the earth.

JUPITER

If the interior of the earth is powered by cold fusion, then what of the other planets? Jupiter, for example, radiates twice as much energy as it receives from the sun. Its output is an incredible million trillion (10^{18}) watts! But where is all this energy coming from?

Jupiter is called a giant gas planet, for, despite its enormous size, it is composed mainly of gases like hydrogen and ammonia. The enormous pressures at its core, however, compress the hydrogen gas into a solid form. The center of Jupiter therefore consists of one of the most bizarre substances in the universe—hydrogen metal, a form of hydrogen that, because of the high pressures involved, is never seen on earth.

Suppose, Steven Jones speculated in a 1986 paper, that deuterium and light hydrogen are fusing within the hydrogen metal at the planet's core. It turns out that the fusion rate need not be very high—absurdly low in fact. If each pair of deuterium nuclei take 10 trillion (10^{13}) years to fuse, then this alone is sufficient to produce all the energy needed. When the enormous density of the core of Jupiter is multiplied by this tiny fusion rate, it produces an amazing 10^{30} fusion reactions each second! This rate of cold fusion would produce exactly the amount of heat—10 trillion watts—that is radiated from the surface of Jupiter.

If cold fusion is a reality, then it will revolutionize our ideas about the earth and planets. Not only may cold fusion prove to be a new source of energy here on earth, but it also will give us a very different view about the universe we live in.

THE SUN

For several decades it has been assumed that the reaction that powers the sun is the hot fusion of protons (hydrogen nuclei), via what is known as the carbon cycle, to produce helium. This means the temperature at the sun's core is hundreds of millions of degrees.

The major difficulty with this assumption is what is called the problem of the missing neutrinos. Neutrinos are elementary particles with no charge or mass; they are produced during the hot-fusion cycles that occur within the sun. As a direct consequence, many neutrinos should be observed streaming out of the sun, some of them hitting the earth. Admittedly, solar neutrinos are detected here on earth, but not enough of them. What has happened to the "missing" neutrinos? With all that heat being pumped out of the sun, far more neutrinos should be observed.

Now cold fusion suggests an alternative explanation. Possibly the temperature at the sun's core is not as hot as everyone previously supposed. Possibly, Paul E. Palmer and Steven Jones suggest, only a percentage of protons inside the sun (the ones moving very quickly) are reacting through hot fusion, while others are reacting through a mechanism that is closer to cold fusion. If the sun's heat is actually being produced by two processes, then this would account for the heat produced while allowing for a lower number of neutrinos. (The neutrinos would not be produced in the cold-fusion reaction.)

COMETS

Another suggestion, which comes from the author of this book, is that an interesting place to look for cold fusion would be in comets. Cold fusion, according to Paul E. Palmer's view, requires water (for the deuterium) plus a mixture of rocks and other materials—one of which may be carbon. Furthermore, fusion appears to occur when a system is stressed away from its normal equilibrium with heat or pressure. This is just the condition of a comet as it approaches the sun, and its chemical composition is ideal.

As the comet begins to heat up and move closer to the sun, it will experience a variety of internal stresses. Possibly under these conditions cold fusion of its water will accelerate, and tritium and helium 3 will be released. Evidence for these two elements could be found by looking at the wealth of data accumulated on Halley's comet.

Paul E. Palmer is pursuing the whole question of geological and planetary cold fusion with great energy. One of his plans is to explore cold-fusion water absorbed in rock and to subject the material to bursts of high pressure and temperature.

Just what is going on?

Steven Jones was driven to experiment with a cold-fusion cell after attempting to reproduce the condition of heat within the earth.

Pons and Fleischmann were following up some curious data on the way hydrogen isotopes form at a palladium electrode. In both cases, it was the particular nature of the metal electrode that held the key to cold fusion.

Assuming that cold fusion is occurring at some level, what exactly is happening deep within the palladium (or titanium) metal? The remainder of this chapter examines just a few of the new ideas and original theories that scientists are using to explain cold fusion. The basic idea is that deuterium atoms must be held close enough for electron tunneling to take over. But to explain exactly how this happens demands the ingenuity and intuition of chemists and physicists to discover ways in which the laws of nature can conspire to encourage cold fusion.

The place to start is with the electrode itself, palladium metal. Although tungsten and even other metals have been used in cold fusion experiments, let us concentrate on palladium.

THE PALLADIUM LATTICE

The first step toward cold fusion occurs at the surface of the palladium itself. Instead of bubbling off as deuterium gas (D_2), the surface of the metal allows the deuterium atoms to remain apart and move into the bulk of the metal.

The drawing of the atomic structure of palladium may look very complicated, but in fact it is made out of series of repeating cubes. The drawing of one of these cubes shows that palladium atoms sit at the corners of the cube with an extra atom in the center of each face.

Figure 6-3

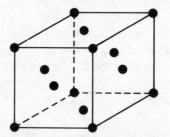

Palladium metal is made up of a series of cubes with a palladium atom at each corner and in the center of every face.

When a deuterium atom enters the metal, it gives up its electron, which is shared democratically with the other electrons of the metal. The positively charged deuterium atom (that is, a bare nucleus consisting of one proton and one neutron) is now free to move through the lattice until it settles down into a comfortable position between the palladium atoms. Finally, when enough deuterium has been pumped into the palladium, every vacant position is occupied, and the lattice is said to be saturated.

Figure 6-4

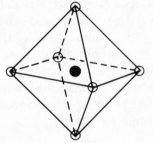

In saturated palladium metal, each deuterium atom (dark circle) is surrounded by six palladium atoms.

But even at this stage, the deuterium atoms are not rigidly held in their positions and can still move. The question is how close they can approach. In deuterium gas, the nuclei are never close enough for quantum tunneling to make any sense. Is there some special property of the palladium that allows them to approach more closely? The answer is "screening."

Screening

Deuterium nuclei are positively charged and repel each other. But the "gas" of negative electrons in palladium is free to move anywhere and tends to congregate around the positive deuterium nuclei. This concentration of negative charges screens off the positive charge on the deuterium nuclei and makes that charge less repulsive. Electrons are like the mediators who come between battling adversaries and reduce the tension between them. Thanks to the screening of their repulsion for each other, the two deuterium nuclei can approach closer than before—close enough to allow quantum tunneling.

Palladium is an ideal metal to show screening. Drs. Chandre

Dharma-wardana and G. C. Aers of the National Research Council of Canada have calculated that it enhances the chance of fusion by a factor as great as 10^{46} (1 followed by forty-six zeros)! Moreover, if the temperature of the metal is raised by 700° C, then this enhancement increases by over a million. Temperature should have a dramatic effect on cold fusion.

But even this electronic screening is not enough to explain cold fusion. Is there a way to push the fusion rate up even higher? The two scientists calculated what would happen if the deuterium nuclei themselves are dislodged from their positions within the metal lattice. Independently of the two Canadian scientists, Dr. K. B. Whaley of the University of California has also shown how the deuterium nuclei themselves will assist in this screening and push up the fusion rate in a dramatic way. The Canadian scientists concluded, "Thus not only observable rates of nuclear fusion, but also economically useful rates should be achievable in these systems."

UTAH FUSION OR BRIGHAM YOUNG FUSION?

One of the most energetic debates on cold fusion has been between the different interpretations of Jones and of Pons and Fleischmann. Steven Jones believes that cold fusion is a low-level effect. At present no detectable heat is being given out, and to push cold fusion to a level where measurable energy is given off would require a lot of luck and some very careful research. Yet from his first research into muon-catalyzed fusion, Jones has been working toward some way of improving cold fusion so that usable energy can be extracted.

Pons and Fleischmann, however, claim that usable energy is already being created in their cell and that it is only a matter of time before that energy can be harnessed for human good. Moreover the source of this energy is cold nuclear fusion. But Jones hotly denies that any nuclear reactions are involved in this heat and suggests that the whole thing is simply some unknown chemical effect, or possibly the release of energy from metal under stress. After all, Pons and Fleischmann spent over 100 hours pumping energy into their cells; maybe some of this energy was later released as heat.

By midsummer 1989 most scientists were coming down on the side of Steven Jones. Pons and Fleischmann fusion was just not believable, and so many laboratories had failed to confirm the Utah findings. As to those excursions of heat, they must be due to some other, nonnuclear effect. The very low-level nuclear fusion claimed by Steven Jones sounded more reasonable, but even in this case, there were critics, scientists who could not reproduce the Brigham Young findings.

The key to settling this whole question will come in two ways. One will be to identify directly the products of a nuclear reaction (for example, helium 4 atoms) and in the right quantities to account for the heat released. The other approach will be to prove that more energy is actually being generated in a Pons and Fleischmann cell than has been put in.

Heat is certainly being given out of a Pons and Fleischmann cell but only after the device has been pumped full of deuterium for over 100 hours. During this time period, electrical energy is being pumped into the system, and electrical energy is still supplied during the running of the cell. The question is whether the energy emitted from the cell as heat exceeds the total energy that has been pumped into the cell. If this is so, then one must conclude that nuclear power is involved. If, however, the net heat released is less than all the power put in, then some other explanation is possible. Essentially the device would be working like a rechargeable battery—with energy pumped in during the charging period and later released as heat.

Only a very careful measurement of the total energy input and output of the cell will satisfy the scientific community. Already some scientists have done some preliminary studies which tend to confirm the Pons and Fleischmann claim. Time will tell if these studies are confirmed.

EXOTIC REACTIONS

Jones and his co-workers at Brigham Young University assumed that deuterium nuclei fuse by the following conventional fusion reactions:

$$\text{Deuterium} + \text{Deuterium} = \text{Helium 3} + \text{Neutron} \qquad (1)$$

$$\text{Deuterium} + \text{Deuterium} = \text{Tritium} + \text{Light Hydrogen} \qquad (2)$$

(The tritium later decays with a twelve-and-a-half-year half-life to produce helium 3.)

The result of reaction (1) is the emission of a neutron. By detecting the number of neutrons emitted from their cell, the Brigham Young researchers had a good idea of how many fusions were taking place.

But Fleischmann and Pons have speculated that only one in every 10 million fusions uses routes (1) or (2). The vast majority involve a new, "exotic" nuclear reaction. In other words, for each neutron that is observed, some 10 million more hidden fusion reactions are giving off heat.

The advantage of an exotic nuclear reaction is that "clean" energy is produced—no nuclear radiation is being given out in the heat-producing fusions. In a regular nuclear fission reactor, hazardous radioisotopes are built up in the fuel rods. Even with high-temperature fusion, enough neutrons are produced to make the tokamak's structure radioactive. But Fleischmann and Pons claim that something very different is happening—a "clean" nuclear reaction, one in which two deuterium nuclei fuse directly to produce a helium 4 nucleus, heat, and no nuclear radiation.

One of the first people to come up with a theory of clean cold fusion was the brilliant MIT theoretician Peter Hagelstein. It was Hagelstein who had first worked out the theory of the x-ray laser used in the "Star Wars" program, but he then left the Lawrence Livermore National Laboratory to work on the more peaceful uses of science.

On hearing about cold fusion, Hagelstein dropped everything and worked night and day on a new theory of clean fusion. In a matter of days, he had written four papers, which were submitted to and accepted by *Physical Review Letters*. At the same time, MIT filed patent applications connected with Hagelstein's analysis. (Later, as befits the bizarre and confused nature of the cold-fusion story, Steve Koonin of the University of California announced that Hagelstein had withdrawn his theory. This rumor was widely circulated and was even carried on Douglas Morrison's "Cold Fusion News" on Bitnet. The story was not, however, true.)

For Hagelstein, the key to clean fusion lies in the palladium itself, which enhances a radiationless form of fusion at the expense of other, "dirty" reactions. Hagelstein's theory is quite complicated and relies on what are called cooperative phenomena. Instead of atoms and electrons all going about their own business, like people in a crowd or bees in a swarm, they begin to cooperate and to move as one. The same sort of thing happens in a superconductor—the electrons in the metal move in a highly coordinated way and are able to flow around any obstacle without resistance.

Hagelstein likewise argued that cold fusion involves the cooperation of many small effects that add up to produce a large fusion rate. In fact, cold fusion has something in common with the laser, in which energy is also released through cooperative processes.

But in Hagelstein's theory, the actual fusion must be instigated by some kind of singular process—a cosmic ray passing through the metal, for example. This even triggers a large number of rapid fusions, which then give their energy directly to the lattice.

Hagelstein's was not the only theory of clean fusion. Charles J. Horowitz of Indiana University suggested that it is the dense electron gas within the palladium metal that makes all the difference. The bunching and screening effect of these electrons around the deuterium nuclei allows the nuclei to approach close enough for quantum tunneling.

But what if one of these electrons is captured in the process? It will be dragged into the fusion reaction and later released:

$$\text{Electron} + \text{Deuterium} + \text{Deuterium} = \text{Helium 4} + \text{Electron (3)}$$

The by-product of this reaction is ordinary helium and no nuclear radiation—with an electron being shot out at high speed and its energy being given to the metal as heat.

Horowitz suggests that the conditions within the palladium conspire to create clean fusion while suppressing other reactions. Clearly reaction (3) would be a highly efficient way of producing heat, since none of the fusion energy escapes as radiation.

One of the major disappointments of the exotic-reaction theory was the failure of John Appleby from Texas A&M University to

ALTERNATIVE THEORIES OF COLD FUSION

A variety of other theories of cold fusion surfaced in the weeks that followed the University of Utah's announcement.

Cosmic Rays

What if naturally occurring muons, generated from cosmic rays, are entering the palladium electrode and sparking off fusion reactions? One scientist speculated that cold fusion worked so well in Utah because of its high elevation—cosmic rays are abundant. The idea was supported by scientists like A. J. McCevoy of the Ecole Polytechnique in Lausanne and by C. T. D. O'Sullivan at the University College of Cork in Ireland.

But James S. Cohen of the Los Alamos National Lab and John D. Davies of the University of Birmingham suggest this effect would be far too small. Experiments performed in deep underground laboratories where the muons and cosmic rays are rare appear to discount this theory.

Quasi Particles

Theoreticians dealing with the properties of metals often describe them in terms of "quasi particles" and theoretical constructs like "heavy electrons." A number of scientists have speculated that these heavy electrons could mimic muons in producing many fusion reactions. But a heavy electron is essentially a theoretical construct, a way of making calculations, and does not literally correspond to a heavy physical electron that could bond two deuterium nuclei into a molecule, as happens with a muon. In fact, the idea of heavy electrons is really a crude way of talking about electron screening.

Other Fusion Reactions

Deuterium may not be the only nucleus that is fusing in the palladium electrode. Lithium is also present in the fuel cell, and one cannot rule out the following reaction:

$$^6Li + D = {}^4He + {}^4He$$

The results at Texas A&M suggest that lithium is critical in the fusion reaction—the excess heat falls off when sodium replaces lithium. But the reason for this effect is not understood.

Fracto-Fusion

Cracks, fractures, and other mechanical defects are found in metal lattices. As the electrically charged particles in the lattice are forced apart, enormously strong electrical fields are generated. These electrical fields accelerate deuterium nuclei to very high speeds across the crack.

This activity resembles a microscopic elementary-particle accelerator or tokamak in which deuterium nuclei are fired at each other at high speeds. The result would be tiny hot spots in the metal, localized regions in which bursts of neutrons would be observed.

This theory has been advanced by a number of scientists and is currently thought to provide the best explanation for cold fusion. The idea is that metal is placed under tremendous stress as it is pumped full of deuterium, and this leads to all kinds of microscopic cracks and fractures. Indeed, scientists involved in cold-fusion research agree that the metallurgy of the electrodes and the metal into which deuterium gas is pumped are critical.

There are speculations that cold fusion is somehow being catalyzed along grain boundaries or microcracks within the metal. Sudden physical changes within the metal's internal structure could be responsible for sudden bursts of neutrons.

Another possibility is that the whole metal, after experiencing great stress during the charging period, begins to release this energy in the form of heat. Or a form of controlled burning of deuterium and oxygen may occur within microscopic cracks. All these theories remain to be tested, and this will require months and even years of careful research.

Rearranging the Nucleus

Peroni Paolo, writing to *Nature* from Rome, suggested that another form of radiationless reaction is possible. When two deuterium nuclei meet in a tokamak at high temperature, they have no time to rearrange their internal structures. But under the very long time scales of cold fusion, the proton on one nucleus could be repelled by the proton on the other. The sequence is shown in the following diagram.

Figure 6-5

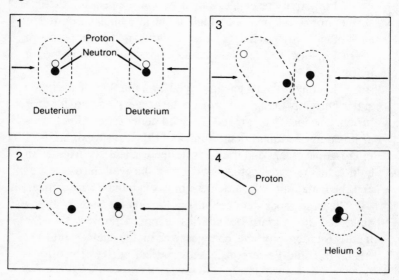

discover any helium 4 in his electrodes. If cold fusion leads to the production of helium 4, then atoms of the substance should be detected within the cell. However, for a variety of technical reasons, helium 4 would be difficult to detect, and its presence would be masked by deuterium. More careful studies must be carried out before it can be definitely said the helium 4 is not appearing in the fusion cells as a by-product of the release of heat.

THE EXOTIC CONCLUSION

- The surface of the palladium metal acts to prevent deuterium bubbling off as gas.

- The internal structure of the metal allows the deuterium nuclei to move with ease and fill up spaces between palladium atoms.

- The gas of electrons acts to screen the nuclei and reduce their mutual repulsion. With a closer approach, quantum tunneling makes fusion more probable.

- Electrons and the metal lattice become involved in and create a clean and radiationless form of fusion reaction.

Theoretical physicists and chemists appear to have no problem in accounting for cold fusion; a host of different theories predict cold fusion at a variety of rates. In particular, some of these theories suggest ways of accelerating and improving the fusion process. However, only time will tell how well they stand up to a closer examination.

In cold-fusion research, experiment and theory will go hand in hand, each helping the other. Theories already suggest that the cold-fusion process should be accelerated by changes in temperature. On the other hand, there are many puzzling aspects to cold-fusion experiments that need the help of a good theory. Good theories suggest new ways in which scientists can look for experimental effects, but the theories must be built on experimental fact, not on rumor and supposition.

Before cold fusion has the hope of becoming commercially significant, a great deal of careful science will be needed. At present it is not even clear if a nuclear reaction is taking place. Untangling the question of cold fusion will take years or even decades. The next chapter will look at some speculations on the cold-fusion world of the future.

Chapter 7
Implications

Within a month of the first cold-fusion announcement, scientists at the University of Utah were talking in terms of scaled-up fusion reactors and cold-fusion systems for use in trucks and automobiles. While Steven Jones at Brigham Young University was cautioning that current attempts at cold fusion were simply a small step on the way toward a long-term goal, Pons and Fleischmann's colleagues were already thinking in terms of the engineering challenges involved in building large-scale power systems.

A number of scientists now agree that there must be something to cold fusion. What is still being hotly debated is the scale of the whole process. There does appear to be evidence of some low levels of fusion, possibly fracto-fusion, in palladium and titanium electrodes. But this is a long way from anything approaching heat production. Some believe that years and even decades of hard work are needed to produce competitive cold-fusion energy, while others maintain that cold fusion simply operates below the margin and can never compete with other forms of energy. The optimists have no time for debate. For them, cold-fusion energy has already arrived, and with each new experiment, the amount of power generated is increased. If the optimists are correct—and that is still a big if—then cold fusion will indeed be the most significant scientific discovery since fire.

But such a cheap and abundant new energy source attacks a

number of vested interests. This could spell trouble. The first to react were the physicists, who felt that their whole understanding of nuclear reactions was being turned upside down—and this was being done not by fellow physicists but by a bunch of chemists! This cold-fusion revolution generated animosity and frustration as some scientists attacked what they felt was a fusion hoax.

But the implications of cold fusion go far beyond the scientist's laboratory. If cheap cold-fusion energy is ever achieved on a large scale, then it will revolutionize not only the energy industry but also the whole web of international politics and trade tied to it. The global effect of this new energy source is so great as to be unpredictable—in fact, we have little in the past to compare it with.

A major new source of energy would have a destabilizing effect nationally and internationally. Already, some scientists have begun to worry about the official reaction to this breakthrough. Research on cold fusion has so far been carried out freely, but some scientists began to worry about whether the U.S. government would suddenly step in. At the time of the Manhattan Project (the building of the U.S. atomic bomb), wide legislation was enacted to control the use and possession of radioactive substances. In essence, the government could clamp down, at a moment's notice, on cold-fusion research within the United States. But in the first months of cold-fusion fever, the exact opposite was happening, with the Department of Energy actively encouraging conferences and research.

Another scare story involved the reaction of power companies. One version, carried on the computer networks, came from "an engineer at a Midwestern power company" who claimed that cold fusion had badly shaken the industry. It is difficult to confirm the accuracy of this story, but the fact that it was being circulated demonstrates the unease that some people were beginning to feel. The power companies had already discussed ways to head off the potential catastrophe of individuals and private interests developing their own energy sources, the engineer reportedly claimed. One approach would be to control the production and possession of deuterium and heavy water, but this looked too difficult to enforce. Another plan was to urge authorities to act on the grounds that the processes involved in cold fusion would be potentially hazardous if they got into too many hands.

But by the time of the Los Alamos conference in late May 1989, nearly every university in the United States and many others throughout the world had an interest in cold fusion. It was simply too late to put the lid on.

The most active group interested in the control of information on cold fusion was not the government or power companies but patent lawyers. The University of Utah was attempting, at all costs, to hold onto the potential of its invention, and that meant having cast-iron patents. The law firms retained by the University of Utah, as well as the patent lawyers operating for other universities and laboratories, were desperately trying to keep the lid on cold fusion.

Patents are complicated things. A drug company, for example, may spend hundreds of millions of dollars and years of research on developing a new drug. But patenting that particular molecule or chemical process is not sufficient. Suppose that a related chemical can do the job just as well or almost as well; it will not be covered by the patent, and a rival company will be free to manufacture it. Whenever a new drug is marketed, drug companies must spend a great deal of effort in researching and patenting the potential of all related molecules and chemical processes. The aim is to own an umbrella of patents that block any competition.

The same is true about the Fleischmann and Pons discovery. It is one thing to patent a Pons and Fleischmann cell, but what about the Frascati discovery in which deuterium gas was pumped into titanium? What if some other metal or new design of fusion cell proved to be as effective, and its use is not included in the Utah patent description? The lawyers had to cover all possible situations.

Scientists who had been unable to duplicate the Pons and Fleischmann result were already beginning to wonder whether there was a secret missing ingredient in the experiment. When Fleischmann and Pons refused to come up with details on request, this only heightened their suspicion. If there really was a magical missing step, this could well be the key to the Utah patent application. Provided that only Fleischmann and Pons knew the secret, their patent would be secure.

But how specific was that magical ingredient? A patent disclosure has to spell out all the details. It is no use talking about "a pinch of chemicals"; the exact composition of the fusion cell has to be spelled out. Moreover, the device has to be shown to work. Some researchers were beginning to wonder whether Fleischmann and Pons really knew what they were doing. It was certainly true that they could get their cells to work most of the time, but did they know enough to make a totally secure patent disclosure?

For the University of Utah, this patent application must have seemed like a Holy Grail. And, like the grail of Arthurian legend, there was always the chance that it would vanish into thin air just as it came within grasp. Too many people were working on cold fusion, and too many speculations were flying around. This is why the lawyers were beginning to be concerned about the wide availability of information on cold fusion—if too many people talked about what was going on, then applying for a patent would be impossible.

THE INFORMATION REVOLUTION

But are these hypothetical governments, power companies, and patent lawyers really living in the present day, or has new technology already passed them by? One of the most exciting things about the cold-fusion story is the speed with which information leaves the laboratory bench and races across the world. Thanks to telephones, fax machines, computer networks, and satellite conferences, a new result spreads across the earth in minutes. A physicist in Italy can learn about a new discovery in a laboratory on the West Coast of the United States even before the American scientist's colleague down the hall. Geographical distances become unimportant, as everyone becomes connected at the speed of light. In such an environment, it is no longer possible to keep the lid on cold fusion; scientific knowledge has become public property.

Some scientists have harshly criticized what is going on and feel that the "due process" of science has broken down. Information being spread electronically, they say, may mean the death of scientific journals and reviewing by scientific peers.

That may be true; the scientific journals and the slow-paced refereeing system may simply be too cumbersome where new innovations are concerned. For the day-to-day slow, incremental progress of science, journals may still survive. But for sudden innovations like cold fusion, high-temperature superconductivity, and the appearance of a supernova, there is growing up an instant process, a sort of electronic peer review system in which ideas surface, are tested out, and are rejected all within the space of hours or days. When a story like cold fusion or high-temperature superconductivity breaks, there is no time to wait for learned journals or academic conferences; every day counts, and scientists have begun to make use of every aspect of the information revolution.

There may well be no going back from this instant propagation of new knowledge. Scientists find this fast process too convenient to renounce, and they will soon learn to evolve new structures for scientific debate. Perhaps only those theories and experiments that are considered valuable in the new scientific marketplace will survive long enough to be printed in the scientific journals. Others will simply die from the electronic networks and be forgotten.

Already there are far too many journals and too many new papers for a scientist to keep on top of his or her field. This electronic filtering process may well serve a valuable function in throwing out poor experiments and half-baked ideas and allowing only those with value to survive into print. In the future, a science may proceed by a sort of Darwinian evolution in which only the fittest ideas survive. It will not matter if a new theory comes from Princeton, Oxford, or some tiny college in the Midwest; the theory will have the same chance of competing for survival.

Of course, there will always be room for good ideas, and these will find their way into print as before. In fact, there are precious few good ideas in science. Most of what is written that is supposed to be new shows little imagination, intuition, or understanding.

With all the modern means of communication that are available, it is no longer possible for any government to exercise total control over a population. Information has become the new global currency, replacing energy, money, and human labor. Information, by its very nature, is distributed and global. Its movement is too fast to catch and hold onto, its structure too subtle to control.

By the end of this century, information will be a totally free commodity, something available to all, no matter where they live in the world. With the freeing of knowledge and its total democratization, the structures of governments and power blocks will crumble.

Cold fusion is one of the new technologies that cannot be kept quiet or controlled. Already too many people know about cold fusion, so there is no way that one company or one nation can ever control the knowledge needed to build a cold-fusion device. If true commercial cold fusion should turn out to exist, then this source will be available to everyone.

THE POLITICS OF ENERGY

As far as the U.S. Department of Energy was concerned, cold fusion was a Good Thing. Beginning with experiments on muon-catalyzed cold fusion, the department had already been giving some modest support to Steven Jones and his group. But when the Fleischmann and Pons story broke, things suddenly looked more promising, and Secretary of Energy James D. Watkins, urged his various laboratories to pursue with vigor research into cold fusion. A special panel was set up to advise the secretary and to produce an interim report by the end of July and a final report in November. A team of scientists visited the University of Utah on June 2 to examine the cold-fusion findings and to report back to the department.

The Los Alamos National Laboratory, along with the Department of Energy, sponsored a workshop on cold fusion on May 23–25, attended by the major players from around the world. Because the meeting was transmitted by satellite throughout the North American continent, it could be picked up live by laboratories and cold-fusion enthusiasts alike. Clearly the U.S. government felt that scientists should be given every encouragement to pool ideas on cold fusion. If this new process could be shown to work at a commercial level, then its implications would be truly revolutionary—a form of cheap and abundant energy that could well push the world into a new era.

The so-called Atomic Age has not lived up to its original expectations—"energy too cheap to meter." If anything, its major impact has been psychological, more to do with nuclear weapons than power stations.

Figure 7-1   ELECTRICAL GENERATING CAPACITY, 1985
(in megawatts)

Country	Thermal	Hydro	Nuclear	Geothermal
United States	533,611	84,986	81,566	1,712
Soviet Union	229,936	61,257	28,100	0
Japan	110,291	34,337	24,686	214
Canada	30,938	57,458	10,889	0
West Germany	69,938	6,668	16,095	3
China	55,700	26,500	0	0
United Kingdom	56,352	4,190	7,064	1
Italy	36,098	17,166	1,273	439
Norway	243	22,991	0	0

Source: United Nations, *Energy Statistics Yearbook*, 1985.

In fact, a more significant revolution over the last decades involved the processing and dissemination of information with high-speed computers, communication satellites, and information networks of all kinds. It is the information revolution, along with robots, microchips, word processors, expert systems, and artificial intelligence, that is changing our world, not atomic reactors.

But the seed of a further revolution had been planted—the cold-fusion revolution. Cold fusion is still a big question mark. But let us indulge in a little science speculation, let us create a science fiction world in which cold fusion is a reality. Imagine a world in which energy is no longer considered an accountable commodity but is free to everyone. What sort of a world would that be? Imagine also an energy revolution in a world that is already in the throes of an information revolution. This adds up to a profound change and a host of possibilities.

In the nineteenth century, available energy was the means by which the human race could exert its control over matter. With energy, raw materials can be extracted, transformed, and transmuted; work can be done; goods can be moved; new materials can be created; new cities can be built. Yet all this was carried out in a hierarchically structured society in which information flowed from near the top of the pyramid to the masses below in the form of orders and directions on how to use all this energy to augment the efforts of their muscles.

Today it becomes possible to leap to a new level of organization, one in which information is freely available to all and can be used to control and shape this free and abundant energy. Provided that this information is no longer concentrated in one country, class, or organization, then the control of energy becomes truly democratic.

CONSERVATION AND COLD FUSION

But not everyone is optimistic about the cold-fusion revolution. Indeed, some of the comments made by the ecology and conservation movements are distinctly negative. Jeremy Rifkin says that the successful harnessing of cold fusion would be "the worst thing that could happen to our planet." His prediction is that more energy would mean more and more people.

Laura Nader has argued that more and cheaper energy would not necessarily mean that people would be better off or that the quality of their life would improve.

Paul Ehrlich, the Stanford University biologist, has said, "the prospect of inexhaustible power from fusion is like giving a machine gun to an idiot child."

Barry Commoner has observed that fusion is bad because people would neglect solar power—a more appropriate energy form for the planet.

One would have guessed that the ecologists, in an age of acid rain and possible climatic change brought about by the greenhouse effect, would have been the first to welcome a nonpolluting form of energy.

It is not difficult to object to certain of their predictions—why, after all, would abundant energy mean more people, when a rise in living standards is generally followed by a fall in the birth rate? But the deeper reason behind their pessimism is not hard to see.

Faced with the possibility of an unlimited source of power, ecologists are naturally concerned with Western society's habit of jumping onto the bandwagon without bothering to see where it is going, and of demanding more and more progress at any price. Throughout the nineteenth century, many areas of Europe and North America underwent an industrial revolution in which they celebrated the power of a few men's minds, and many other men, women, and children's bodies, to transform the world around them. The earth was a planet for the taking, and the movers and shakers of Western society reveled in their power. Coal and ores were mined; metals were created in blast furnaces; iron ships, trains, and bridges were built; and the New Age was celebrated in such creations as the Eiffel Tower and the Crystal Palace.

Progress was a Good Thing, and more necessarily meant better. A constant increase in productivity became the goal of civilized nations, and novelty was the keyword. Even into the 1930s Ezra Pound cried, "Make It New." Even if the great British Empire's sun was setting, the stars of the United States were rising above its skyscrapers and iron railways.

Only relatively recently has this triumph of the new been questioned. Within the past few decades, ecologists have shown us what we have really been doing to our planet, and conservationists have convinced us that the earth's resources are finite.

Today there is an increasing tendency to think of the earth as Gaia—as if our planet with its resources, earth, water, air, and living inhabitants were a living organism. Some Western thinkers have turned to the philosophies of the East or to the wisdom of the original indigenous peoples of the Americas to seek new ways of living together in harmony on the planet.

Many thinkers have rejected progress and increase in favor of conservation and technologies that are appropriate to each situation. Megaprojects are frowned upon, new endeavors must be assessed in terms of their effects on the planet, quality of life is substituted for standard of living, and economists are being forced to cost the value of clean air and water when any new project is undertaken.

A number of serious thinkers are calling for a pause in the effort being made by the human race. They ask for a pause in endless progress, a pause in the proliferation of genetic engineering, a pause in new megaprojects, a pause in the building of nuclear power stations, a pause in the spiral of aggression and conflict, a pause in which the whole globe can catch its breath. Within this pause, there would be a time to rethink what societies are doing and to take stock of the responsibilities of each individual. It would be a pause that empowers everyone and gives each person a new sensitivity to the complex world he or she lives in.

Into this evolving consciousness of the problems that face us, cold fusion has burst. It is only natural that the ecologists and conservationists should worry that yet again the human race will lose its sense of balance and, intoxicated with the promise of abundant energy, will lose touch with everything that has been gained in the last decades. How easy it would be to say, "With clean and abundant energy, we can forget about acid rain and the greenhouse effect. We can forget about the balance of payments. We can forget about limiting growth. We can forget about the gap between rich and poor. Because all these problems can now be solved."

THE COLD-FUSION POTENTIAL

Steven Jones may not agree with Pons and Fleischmann when it comes to the mechanics of cold fusion, but on one topic they are unanimous: the need for a clean, safe, and abundant source of power. Moreover, they agree on the hope that this power could eventually come from cold fusion. It is something that could transform not only society but the whole planet.

But just what is so special about cold fusion? And why, if it works, should its impact be so remarkable?

A good way to understand the cold fusion revolution is to compare it with nuclear fission. The world's first chain reaction was achieved in an atomic pile built at the University of Chicago on December 2, 1942.* To construct his reactor, Enrico Fermi had to use something as big as the university squash court, which was piled

---

*In fact, Fermi's atomic pile was not the first to be built in the world. A Canadian scientist, Dr. George Laurence, was also busy building a nuclear reactor in the basement of the National Research Council of Canada, Ottawa. Laurence piled up paper coffee bags filled with uranium oxide and interspersed with coke, which was to act as the moderator. But Laurence's atomic pile never became critical—in fact, the materials he was using contained so many impurities that a chain reaction would have been impossible. The Canadian failure resonates well with similar failures to duplicate cold fusion today. It is not just a matter of putting all the ingredients together in the right way, but a number of other critical factors have to be just right. It is only later, when the full theory of nuclear fission (or cold fusion) is properly understood, that it becomes a straightforward matter to duplicate the process.

almost to the ceiling with uranium interspersed with graphite blocks. An atomic pile, even the very first prototype, is a large object, and present nuclear reactors are big and costly. Even a small-scale self-contained reactor such as Slowpoke, developed by Atomic Energy of Canada Ltd., requires special excavation for its installation. Atomic energy does not lend itself to the small scale.

Running tokamaks and inertial confinement systems requires all the engineering and scientific facilities of a major research laboratory. It is not sufficient to built a doughnut-shaped ring, for the tokamak must be supported with such major pieces of equipment as superconducting magnets, microwave heating (rf) devices, particle beam heaters, and a host of electronics and control systems.

Nuclear fission and high-temperature nuclear fusion are both major, capital-intensive businesses. They call upon a number of support industries, along with scientists, engineers, and a variety of specialized companies.

The conventional route to nuclear power requires a sophisticated infrastructure and major funding, which not all countries possess. Nuclear power can only be operated at the level of governments of major corporations. This means that nuclear fission tends to become centralized and controlled by a limited number of vested interests. When high-temperature fusion is finally harnessed, the science and engineering involved will be even more costly and specialized. There is really no such thing as "appropriate technology" when it comes to conventional nuclear power.*

Contrast this with cold fusion, speculate what may happen if this technology really fulfills its early promise. Scientists like Steven Jones may believe that the possibility of extracting energy from cold fusion is not yet within our grasp and may never be practicable. Many scientists agree with this position. But just suppose, for the sake of speculation, that cold fusion should eventually work out. What would be its implications?

---

*Atomic Energy of Canada is, however, attempting a scaled-down reactor based on its Slowpoke design. It would be used for heating offices and hospitals and for producing small amounts of electricity. Nevertheless, Slowpoke is at present a large and expensive piece of equipment.

The first attempts to observe fusion certainly were not achieved in a gymnasium-sized reactor or after spending tens of billions of dollars, but took place on a laboratory bench and with a modest budget. The scale on which cold fusion works is totally different from that of other forms of nuclear energy.

Its most exciting aspect is its potential to be scaled to an appropriate size and to provide energy for a wide variety of applications. All indications are that cold fusion will not be confined to a single application or limited to a particular scale.

At present cold-fusion experiments are done in small reaction cells. But already scientists at the University of Utah are building a scaled-up reactor, capable of producing energy at the rate needed to power a light bulb. The next step would be a reactor with enough energy to run a microwave oven. By July 8, 1989, Pons had claimed a water heater powered by cold fusion.

At each stage, new problems must be faced, and engineering difficulties ironed out. However, if the technology works out, then cold-fusion cells will be built in a variety of sizes and powers. There will be no need to spend millions or even billions of dollars on a cold-fusion reactor or to surround it with a building full of elaborate equipment. Cold-fusion cells will probably stand alone. They will be cheap to build and, once running, relatively simple to operate.

Imagine a cold-fusion installation in the basement of your home, providing heat in winter and warm water for showers. One day there may even be a cold-fusion cell powering your automobile. You'll only have to top it with heavy water!

Hospitals, offices, and small businesses will use larger cold-fusion cells for their energy needs: heating and supplying electricity for lighting and small machinery. Remote communities and farms will also be powered by cold-fusion reactors. In the Third World, technicians will be trained to operate and repair cold-fusion cells so that energy can be generated locally in each village and community. Finally there may be banks of reactors to power a city or major industry.

At this point, doubts arise about the safety of cold fusion. Cold-fusion devices capable of generating tens or even hundreds of megawatts pose serious difficulties. If they are extensions of present devices, involving deuterium gas compressed into metal rods or pumped in by electrolysis, an enormous quantity of deuterium would be contained within a restricted area. Suppose that one of these devices should become unstable and release its energy and its explosive gas in a sudden burst. A number of laboratories, including the University of Utah, have experienced sudden detonations while running electrolytic cells. Suppose that such an event should occur with a rod several inches or feet thick and packed full of deuterium. The result would be devastating. Cold fusion on a very large scale could pose as many problems as other energy sources.

But perhaps fusion energy will be decentralized and available at levels appropriate to each need. The only requirement will be heavy-water extraction plants in each country to supply the fusion fuel to customers. The fusion reactors themselves would be manufactured by a variety of companies in as many different types and trademarks as there are models of automobiles today. Since cold fusion would not be capital-intensive and would be available at a variety of scales, there would be niches in the market for individual manufacturers who produce their own customized reactors—cold fusion would not be swallowed up in monopolies.

In short, cold fusion promises much and may well provide an ideal solution to many of the problems that face our world today. With cold fusion, energy may be plentiful yet scaled to appropriate needs.

Furthermore, energy may not be the only benefit of cold fusion. Steven Jones has pointed out that one of the characteristics of cold fusion is the emission of neutrons. As of early June 1989, neutrons were coming out in bursts of over 100 from metal that had been saturated in helium gas. Moreover, these neutrons all had exactly the same energy.

If the number of neutrons could be increased considerably, then it would be possible to use cold fusion as a source of neutrons. This could be valuable in, for example, cancer therapy, particularly as the neutrons would have a known energy and could be directed precisely toward a tumor.

There are, of course, other theories of cold fusion which suggest that energy is released without any nuclear radiation. Utah fusion, if it exists, would not be appropriate as a source of neutrons.

THE ECONOMICS OF ENERGY

Ask an economist what money is, and he or she may reply, "Money is what money does." In other words, money, as green pieces of paper or bits of information on a stock exchange computer network, has no value in itself. The true meaning of money lies in what it can do and how its accumulation and flow affect such things as world trade, employment, the ability to acquire and sell goods, and the standard of living of one country as compared with another. The complex dynamics of money as it moves across our globe drives the activity of governments and international corporations as well as individuals, whether they be peasant farmers in Ecuador or multimillionaires in the United States.

In some ways, energy is similar to money. It is constantly on the move, accumulating in some areas and being actively exchanged in others. Indeed, it is instructive to define energy by saying, "Energy is what energy does."

But just what can energy do? Among other things, energy can:

- Cook food

- Heat homes

- Desalinate water

- Mine and process raw materials

- Turn ores into metals

- Turn metals into automobiles

- Power industry

- Produce new materials

- Transport food and other goods

- Excavate mountains and build houses

- Power communications networks

- Produce fertilizers

- Power the petrochemical industry and create drugs, plastics, pesticides, and herbicides

- Increase a person's standard of living

But energy in the forms we now know also has negative aspects:

- Megaprojects like hydrodams cause adverse geological, social, and ecological impact.

- Hydrolines and pipelines stretch across the land.

- Heavy capital investments are required in centralized plants.

- Pollution and health hazards are generated along with energy.

- Power plants bring about ecological changes such as the greenhouse effect (global climatic change) and acid rain.

- As a by-product, increased industrialization can create ecological and social damage.

- Energy stimulates the demand for a continued increase in the standard of living, which leads to a spiraling demand for ever more energy.

- Power plants are a drain on finite energy resources.

- Oil is being burned for energy, thus diverting it from other potential uses such as the production of chemicals and fertilizers.

- Arable land is used to produce biomass for use as fuel and not to provide food.

- Spiraling energy costs divert funds from other important uses.

- An imbalance in the distribution of energy resources leads to international tensions.

- The use of energy increases the separation between the haves and the have-nots.

The energy equation is a complex one, and cold fusion is going to be a new and powerful variable. Already the world's energy situation gives cause for concern.

THE ENERGY BALANCE SHEET
Biomass

For its energy source, half the world depends on biomass— wood, dung, straw, and charcoal. But 70 percent of those relying on biomass are already experiencing a scarcity—in some areas up to 300 days are taken up in each year gathering fuel for a family. By 2000, 3 billion people will be experiencing scarcity.

Oil/Gas

The world supply of oil is estimated to last for thirty years, coal well into the twenty-first century. But oil has other and more valuable uses than simply being an energy source, since it is the starting point for a variety of chemicals. In addition, burning oil, gas, and coal contributes to the greenhouse effect and to acid rain.

Hydropower

One-quarter of all electricity is generated by hydroelectric power, with two-thirds of this occurring in the developed nations. But electricity is only a small portion of the world's energy pie. Hydropower may appear clean and unpolluting, yet megaprojects like dams cause considerable ecological change and can be disruptive to rural or native population. Previous thinking had been to push for megaprojects, but a new attitude calls for "small hydro"—projects that are more appropriate for the environment.

NEW ENERGY CURRENCIES

Energy, like money, comes in different currencies, some of which—like coal, oil and gas—represent finite "currency reserves." Others, like tidal, wind, hydro, and solar power, are continuously available but are confined to particular areas. Some forms of energy, like oil, are concentrated in certain parts of the world, making these countries energy-rich and powerful.

Figure 7-2

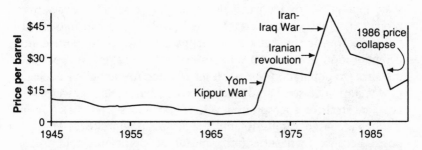

World crude oil prices (based on 1986 $) fluctuate with world politics

Coal, gas, and oil can be easily stored, while other energy forms, like electricity, must be used as soon as they are produced. Some forms, like gas and electricity, are easy to transport, while others, like coal, are expensive to move. Some forms of energy, like electricity, can be used for a wide variety of tasks—heating, lighting, powering machinery, and operating telecommunications systems. Others, like coal, can be used only for a limited range of tasks or must be converted into a form, like electricity, that is more useful.

Not only does energy come in a variety of forms, but these forms can be transformed one into the other as well. The chemical energy trapped within coal can be converted to thermal energy by burning. Thermal energy creates steam to drive a turbine—mechanical energy—and to produce electrical energy. In turn, electrical energy could be used to electrolyze water, producing oxygen and hydrogen (back to chemical energy).

Moving energy among its various forms is like changing dollars into marks, marks into yen, yen into pounds, and pounds back into dollars. In each case, a certain exchange rate is paid, and a penalty is extracted by that "bank" charge called entropy.

In short, what the world needs is not "more energy" or "cheaper energy" but energy that can be made available exactly where it is wanted and in a form that is appropriate for the specific tasks required.

This is exactly why the potential of cold fusion is so exciting, for electrical energy and heat can actually be created on site and at a scale appropriate to each particular need. The great James Bay power project in Quebec, for example, may produce a great amount of hydroelectric power, but no one happens to live in its immediate neighborhood. Instead, all this power has to be shipped over vast distances to population centers. It may be possible to build a nuclear power station close to a population center, but nuclear power does not lend itself to appropriate levels of technology. Nuclear power stations are large, capital-intensive, and not appropriate for all purposes. Because it will suffer from none of these problems, cold fusion looks ideal. The raw material—water—is available to any country. All that needs to be done is to divert a small percentage of cold-fusion energy into the extraction of heavy water from light water. At present Fleischmann and Pons employ expensive palladium electrodes, but some other experiments suggest that the cheaper and more abundant metals could be used.

Unlike a nuclear power station or a tokamak, a cold-fusion cell needs no elaborate technology to set up, simply the levels of engineering available to any country. And all indications are that cold fusion *is* a scalable technology. This means that cold-fusion cells can be manufactured at low cost and in a size that is appropriate for particular needs. A small community in Northern Canada, for example, could use a fusion cell to generate heat and electricity for lighting and to run light machinery. Small villages anywhere in the world could gather their power from fusion cells.

And what about the city? In the initial stages of a cold-fusion revolution, the city's power would still come from a variety of conventional sources. Where hydroelectricity is cheap, it will be used to supply a basic demand. But each city could reduce and level off its demand by using a bank of fusion cells.

Hydroelectric power requires a costly power transmission system that threads across the countryside to bring the energy of a dam or waterfall to a distant population. Those already in place will probably continue to be used. But with cold fusion, there would be no need to build massive centralized power stations. Rather, each town and community would produce all the cold-fusion power it needs at just a few locations. In the first decades of fusion power, no more big transmission towers would be built, although small electrical distribution systems may survive in the city.

And what of small fusion cells for home heating and fusion cells in every car? These are distinct possibilities. In the end, imagination is limited only by economics and the ability of engineers to overcome technical problems.

SOME POSSIBLE SCENARIOS

From the perspective of just several months into cold fusion, a number of scenarios are emerging.

Fusion Power in the Hands of a Multinational Corporation

One scenario is that of fusion power falling into the hands of one multinational corporation. Under this scenario, cold fusion becomes associated with the wealth and a political power never before seen on earth. The chances of this happening are remote, since knowledge of fusion power has already spread too wide, and the fusion process does not call for advanced technology. It is unlikely that a single company will ever hold an exclusive blanket patent for cold fusion.

Fusion Power Controlled by the Developed Nations

High-temperature fusion (tokamak fusion) is enormously expensive. If a breakthrough in this form of energy is ever made, then the technology and scientific knowledge will initially be owned and controlled by the developed nations. If such new power is not used wisely, then it will lead to an ever-accelerating gap between the rich and poor nations.

This is unlikely to happen with cold fusion. Cold-fusion cells are not capital-intensive, nor is the fuel rare or unavailable. The required technological infrastructure is attainable by most countries in the world. Knowledge of cold fusion processes is already being widely disseminated. There would be nothing to stop even the poorest country from using cold-fusion power.

Fusion Power Controlled by Governments or National Companies

Fusion power could be controlled by governments or national companies. This is the way the energy has been handled and distributed in the past. But it may be difficult to centralize and control fusion cells when they are being manufactured by a variety of companies, large and small. While the original fusion processes may be heavily patented, it is possible that a number of other approaches will be developed; already cold fusion occurs with electrolysis and with compressed gases in metals, and in metals other than palladium. The single common denominator of cold fusion is heavy water—and its source, ordinary water, is beyond control. Therefore, chances are that cold fusion will escape the control of governments and major companies.

Cold Fusion Abundantly Available

The most exciting scenario is one in which cold fusion is the exclusive property of no one country or corporation. Cold-fusion devices are as available as automobiles and come in as many different models and colors. There are cold-fusion installations for big factories, cold-fusion power stations in every city. Maybe a cold-fusion plant down on the farm and a cold-fusion heater in every basement, cold-fusion-powered ships, boats, trains, and automobiles.

What would be the timetable for this scenario of democratic fusion energy? It will probably involve a number of phases.

In Phase 1, the infrastructures of conventional energy sources still dominate in the West. But cold fusion is used to augment existing energy forms. Small-scale cold-fusion cells are used in remote communities. People begin to buy cold-fusion cells to heat their homes. The first cold-fusion automobiles appear. In the Third World, cold fusion begins to make important inroads.

In Phase 2, conventional power stations are phased out. Popular pressure increases to restrict energy forms that release acid rain and add to the greenhouse effect. A wide variety of improved cold-fusion systems appear on the market. Many companies and industries are powered by cold fusion. The price of oil and gas continues to drop, and automobiles powered by alternative forms compete with the conventional engines. But with cheap gasoline, some people hold on to their conventional automobiles. The oil-producing nations are faced with a major crisis.

In Phase 3, cold fusion is used in most countries. It is also being used to generate hydrogen as a fuel, and the cold-fusion/hydrogen economy begins. Hydrogen becomes the starting point for a number of clean, nonpolluting fuels, and a variety of new engines are developed to burn these fuels. Oil and gas are no longer burned but become the starting point in the production of a variety of raw materials. There has been a general increase in the standard of living in many of the third-world nations. The Middle East asks for economic aid, and third-world nations announce a variety of long-term goals.

In Phase 4, high-temperature fusion has been successfully developed and competes with cold fusion. Large-scale high-temperature fusion devices are used to produce the great amounts of energy for major cities and specialized enterprises. Cold fusion is considered more appropriate for other tasks.

There has been a wide-scale increase in the standard of living across the world. The cost of fertilizers continues to fall. A number of irrigation projects are under way. Agricultural productivity increases in many nations. The cost of transporting food and other goods continues to fall. Carbon dioxide levels in the atmosphere stabilize, acid rain levels drop, and a number of United States lakes show signs of recovery.

High-temperature fusion has now been commercially harnessed and is used where high power outputs are required. A variety of other technologies are also developing at an accelerating rate. Computing devices are faster and smaller, being based on optical electronics. The contents of the Library of Congress are available on an optical device for a few dollars. A variety of highly sophisticated processing units have already been implanted into the human body, some of these being organically based. Novel materials are designed for specific tasks. Many of these are organic in nature and are grown biologically.

With an increased worldwide standard of living and a better understanding of the immune system, there has been an impressive improvement in general health around the world. The incidence of many degenerative diseases—cardiovascular illness, cancer, aging— has begun to drop. Many infectious diseases have been eliminated from the Third World.

A variety of megaprojects is now proposed:

- Large-scale communities to be created on the moon

- Human expeditions to the outer planets

- Factories to be built under the Pacific Ocean near the tectonic plates for the extraction of new materials.

These projects meet with a mixed reception.

There are plans to divert rivers and to effect climatic change in a number of areas.

The power of governments and corporations continues to erode as the power base moves toward citizen groups, many of which stretch across national boundaries.

But social critics are concerned that, despite the rise in the general standard of living, outstanding problems remain. The whole issue of conflict and violence is still to be resolved. Despite continued research in pharmacology, there has only been a marginal improvement in the treatment of neurotic and psychotic disorders. Crime rates have fallen but still give room for concern. Some ecologists are concerned that new and powerful technology is being used unwisely.

THE BRUNDTLAND REPORT

In the past, the developed nations have generally welcomed continued progress and development. Increasingly, however, many world leaders are concerned that this cannot go on, for our planet is a closed and finite system, and we must learn to live within its limitations. Limitless growth will eventually exhaust our natural resources, create irreversible environmental damage, and increase the gap between the developed and the developing nations.

These concerns were addressed by a worldwide study entitled "Our Common Future" led by Gro Harlem Brundtland, prime minister of Norway. The Brundtland commission investigated many interlocking concerns that were identified as the major issues facing our planet. Energy was one of these.

Current projections of energy use vary widely. By the year 2020 in a world of 8 billion to 10 billion people, the projected increase in energy use ranges from a modest 10 percent to an increase as great as 300 percent.

The Brundtland report calls for what it terms "sustainable development," in which the developed nations voluntarily reduce their demands on the planet by using more efficient processes, eliminating waste, and conserving energy. This will not necessarily reduce our standard of living; indeed, some futurists argue that in a conserver society, the quality of life will improve as the environment is restored. The Brundtland report also allows for the Third World to experience its own period of development—a 6 percent increase in energy consumption by the developed nations, with 3 percent for the world as a whole. Features of this plan include more efficient manufacture, energy-efficient housing, conservation, and increased use of solar, wind, tidal, and geothermal power.

Sustainable development is the wave of the future—although the details of this global plan must be worked out in greater detail. The aim is to reduce our impact on the planet and, at the same time, to reduce the gap between rich and poor nations and to reduce one more of the destabilizing factors that threaten all of us.

It is into this new world view of energy and sensitivity to the planet that cold fusion has burst. Its implications are difficult to judge. In essence, cold fusion suggests a very different scenario, a form of energy that may be relatively clean and safe and that, without the need for a sophisticated technological infrastructure, will be available to all countries, rich and poor.

Of course, cold fusion is still a mystery. No one knows whether it will ever be scaled up to a commercial level. Indeed, there is considerable controversy over whether this process will ever provide usable energy. If cold fusion does pan out, then it must be used wisely and within a new context in which society is sensitive to the debt it owes to the planet and to future generations.

Chapter 8
Conclusions

The press conference called by Professors Pons and Fleischmann on March 23, 1989, together with the subsequent announcement from Brigham Young University, has implications that will stretch far into the future. A totally new range of phenomena has been revealed to the scientific community, information that will be puzzled about for many years to come. In addition, the press conference has shown scientists a new way of working together, unwelcome to many, in which the "due process" of science has been bypassed in favor of instant communications, press releases, and what could be called, for want of a better word, electronic communications.

Also important are the long-term implications of these discoveries. These are what may ultimately bring considerable benefit to society.

To understand what is happening in those electrolytic cells and that metal pumped full of deuterium will require major coordinated research involving nuclear and solid-state physicists, chemists, and materials scientists. That effort will be an integrated program that is rare in science today. Such a project will require major funding, but in addition to studying cold fusion, it will produce many valuable and unpredictable spin-offs as scientists develop their understanding of the new processes involved. All this will be of ultimate benefit to society.

In the first stage, researchers will devote their attention to

173

improving the reproductibility of phenomena such as excursions of heat and sudden bursts of neutrons. They will ask major questions, such as: Is the heat emitted from a Pons and Fleischmann cell the direct result of a new form of nuclear fusion, or does it occur as the result of some as yet undiscovered physical or chemical process?

Four different scenarios or outcomes are possible:

1. Pons and Fleischmann have discovered controlled nuclear fusion, and practical applications are around the corner.

2. The energy released within the Pons and Fleischmann cell has nothing to do with nuclear process and is of, for example, chemical origin.

3. Nuclear fusion does indeed take place, but at the extremely low levels detected by Jones at Brigham Young. The phenomenon is of considerable scientific and geological importance but has no immediate relevance as a new energy source.

4. Cold fusion does not occur under any circumstances.

COLD FUSION

Current scientific consensus is strongly against the first scenario. Indeed, information leaking from the Department of Energy's committee on cold fusion suggests that committee members are very negative about the possibility and even have doubts about the low-level fusion detected by Steven Jones.

The supporters of Pons and Fleischmann have not appeared to be dismayed. According to Stanley Pons, the committee was biased from the start. "I think they were mandated to come forth with a negative conclusion," he is reported as saying. "I don't understand how a panel composed of such people was formed." According to Texas A&M chemist John Bockris, it was a "killer commission."

As of late summer 1989 the state of Utah and University of Utah still appeared to be going ahead with their Fusion Research Center. The state had released $5 million to the university which was treating this funding as seed money for cooperative research projects and corporate participation. There was even talk of General Electric becoming involved in a collaborative effort. With tritium being detected in the fusion cells, Pons and Fleischmann believed that cold fusion had been confirmed, even if a number of questions remained about the amount of heat emitted.

But if nuclear fusion is really taking place in these test tubes, then both the University of Utah and Brigham Young University are seeing the same phenomena but at vastly different scales. Why should this be? Unraveling this mystery will require a full understanding of the various nuclear processes involved and the discovery of how they can be enhanced and developed commercially. This will need a major long-term scientific effort on a national scale and one that will be duplicated in many countries. Not only basic science will be involved, but also all the questions and difficulties required to scale up such a process to the engineering level.

CHEMICAL HEAT

Many physicists, like Nobel Prize winner John Robert Schrieffer, believe that cold fusion, of the type seen by Steven Jones and others, is a totally separate phenomenon from the heat production observed by Pons and Fleischmann. Their conclusion is that Pons and Fleischmann have misinterpreted their data and that this heat is being produced by some new, nonnuclear processes that are occurring in the electrode. When all the heat that has been pumped into the system, both electrically and chemically, is added up, it will be found to exceed the heat given out. In other words, there will be no net generation of heat, no creation of energy.

But this does not mean that the Pons and Fleischmann effect is totally useless. Admittedly it may not be the key to nuclear fusion, but it could be a new and useful way of storing energy—a novel kind of chemical or physical battery, for example.

If this is true, then chemists and physicists must enter an area of new knowledge. They must discover the mechanisms of a new and unknown phenomenon that is capable of generating large excursions of heat from a palladium electrode that has been saturated with deuterium. A variety of physics and chemistry pathways and theories will have to be explored. It will require a detailed study of the process of electrolysis, along with an understanding of the physical, metallurgical, and chemical changes that are taking place in the electrode.

The spin-offs of such a study could be enormous. They will give scientists a new insight into electrochemical processes that form the basis of, for example, batteries that are found in everything from electronic watches and laptop computers to automobiles and space shuttles. Moreover, the heat itself will have valuable applications in new forms of batteries and power systems.

Even more exciting will be the new knowledge to be gained about metal structures and their behavior under the stress. Certainly something is going on in the palladium electrodes that fifty years of solid-state physics had not anticipated. New knowledge in this area is bound to be significant, since it is the science of the solid state that is responsible for many advances in electronics and the development of new materials and alloys.

LOW-LEVEL COLD FUSION

The third scenario in which Pons and Fleischmann experiments do not have a nuclear origin still leaves the whole question of the low-level nuclear fusion discovered by Steven Jones and his group. Although this is still a controversial field, a number of careful and respected scientists are willing to entertain the possibility of low-level cold fusion taking place in metals—as a result of either electron and nuclear screening or "fracto-fusion."

What lies ahead is a major research project devoted to understanding what sorts of nuclear reactions are going on in electrolytic cells and in metals pumped with deuterium gas. Of course, usable energy may never be obtained by such processes, but immediate practical spin-offs are not the only motive for doing scientific research. The ultimate goal will be new knowledge and the discovery of ways to enhance the process to the point where possibly, one day, useful energy can be extracted. Today we have a new path toward nuclear fusion, one that was not available before March 1989.

In the process of getting to understand cold fusion, there is much to be learned about metals under stress and the various reactions that take place within them. Again, there will be many scientific spin-offs. Steven Jones has already suggested one of them: the possibility of a source of neutrons that could be used in the medical treatment of cancer. Another major implication of cold fusion is that it may actually be taking place under our feet and in the planets and other astronomical bodies. This by itself is going to revolutionize a lot of what we know about geology and astronomy.

Cold fusion, whatever it is, still looks like an exciting new field, a field that is filled with question marks. And no one knows what science will discover as it attempts to answer them.

One of the discoverers of cold fusion, Steven Jones, does not hold out much hope to extracting useful energy from this new method. His scientific career has been devoted to discovering a new way of harnessing fusion energy on earth, and he is willing to admit that, from a practical point of view, this is just another dead end. But he is also willing to entertain the speculation of his colleague Johann Rafelski.

Rafelski has pointed out that cold fusion is in its infancy and compares it with the very first experiments on electricity carried out in the sixteenth and seventeenth centuries. These were done on what looked like a range of very different phenomena, none of them looking particularly attractive from a practical point of view. But today this same electricity dominates our world. It is the major source of power in our lives and does everything from heating and lighting our house to powering our computer, television, and telephone. It would be hard to think of a modern world without electricity. Electricity is ubiquitous and used in a wide variety of ways.

But then who would have anticipated this dominating power of electricity when, 200 years ago, scientists first observed a curious new phenomenon associated with metals placed in a solution of various chemicals? The first manifestations of electricity were poorly understood. In fact, it took two centuries to integrate all the different phenomena that are electricity and to learn how to control them. Understanding electricity meant studying static electricity, lightning bolts, the twitching in a dead frog's leg, the mysterious and invisible "fluid" that flowed from metals immersed in certain chemicals, the effects produced by moving an electrical magnet. All were connected, all were part of that one phenomenon we have come to know as electricity.

Cold fusion may be simply a scientific aberration, a flash in the pan to place beside polywater and N-rays. On the other hand, there could be something to it. Those excursions of heat, bursts of neutrons, muon-catalyzed events, radioactive decay products observed in volcanoes and heat plumes may one day integrate into something that is totally new and unexpected.

Cold fusion is a prime example of how science evolves. Not every new pathway is fruitful. Yet when a new phenomenon is discovered, no one can predict how it will turn out. Nature is unpredictable, and science cannot be legislated. Today, only a short time after that dramatic press conference in Salt Lake City, scientists are still playing around with something they do not fully understand. Cold fusion may lead nowhere. Or we may be standing on the threshold of something extraordinary.

Afterword:
Into the Second Year

On March 23, 1990, a radio station in Salt Lake City telephoned me for an interview. "Cold fusion is a year old today," the announcer said, "but just what is happening? Does cold fusion really work?"

How was I going to answer? While a great deal has happened since Pons and Fleischmann first called their press conference, the mystery remains. Could nuclear fusion really take place in a test tube? How does a cold-fusion cell really work? Will energy from cold fusion ever become a commercial possibility?

A year later, the answers to these questions remain ambiguous. While cold fusion has its skeptics and its true believers, the majority of scientists still don't know what to think about the phenomenon. But at least, for those who have studied the available data on cold fusion, one thing is starting to become clear: something curious is definitely happening. It may or may not be cold fusion, but whatever it is can no longer be explained away as bad experimentation or pure fakery. Again and again, careful experiments have demonstrated that something unexplained is taking place in these fusion cells. The problem is that the effect simply is not reproducible from one cell to the next, or from one experimenter to another. Science does not like to deal in mysteries; nevertheless, cold fusion shows no signs of going away.

The months that followed the initial burst of cold-fusion fever, during which hasty reports of confirmations or refutations seemed to be coming in from all over the world, have been devoted to more careful investigations. The main highlights of this period, which are

179

shown in Figure A-1, are a mixture of triumphs and disappointments for the aficionados of cold fusion.

Figure A-1

MILESTONES IN COLD FUSION
1989

March 23	Stanley Pons and Martin Fleischmann hold a press conference in Salt Lake City to announce the discovery of cold fusion.
March 24	A press release from Brigham Young University announces the independent discovery of cold fusion by Steven Jones.
March 31	Martin Fleischmann explains cold fusion to a group of physicists at the European Center for Nuclear Research (CERN) near Geneva, Switzerland. He is warmly received.
April	In the weeks that follow, a series of press releases, faxed reports, electronic-mail messages, and general rumors from laboratories all over the world appear to confirm cold fusion. But soon these give way to negative results in which research groups claim to have seen nothing.
April 7	In a special session, the State of Utah passes the Fusion/Energy Technology Act. Governor Norm Bangerter asks the state for $5 million in research funds.
April 12	Pons delivers a talk titled "Nuclear Fusion in a Test Tube?" to the American Chemical Society in Dallas. He is so mobbed by reporters that he is forced to change his hotel.
April 17	Stanley Pons announces that a fusion reaction has been sustained for 800 hours in one of his cells.
April 18	An Italian group from Frascati announce a different sort of cold fusion, one in which deuterium gas is pumped into titanium metal at high pressure.
April 26	Pons, Fleischmann, and Jones appear before the House Science, Space, and Technology Committee. University of Utah president Chase N. Peterson recommends the establishment of a national cold-fusion center with an initial capital outlay of $100 million. He asks the federal government for $25 million in seed money.
April 27	The cold-fusion paper by Steven Jones and his Brigham Young group appears in *Nature*.

May 1–2	Special late-night sessions on cold fusion are held at the spring meeting of the American Physical Society in Baltimore. Steven Jones is well received, but the physicists present are highly skeptical of Pons and Fleischmann's claims.
May 8	Cold fusion is greeted positively at the annual meeting of the American Electrochemical Society, held in San Diego.
May 23–25	A workshop on cold fusion held at Los Alamos National Laboratory and organized by the Department of Energy attracts scientists from all over the world and is carried by satellite across North America. Pons and Fleischmann do not attend. Negative and positive results are presented, and it is difficult to reach any definitive conclusion about cold fusion.
June 15	At a press conference at Harwell, UK scientists announce that, following extensive experimentation, no evidence of cold nuclear fusion has been established. They are abandoning their investigations.
July 12	An interim report by the Department of Energy appears: "The experiments reported to date do not present convincing evidence that useful sources of energy will result from the phenomenon attributed to cold fusion."
August	A National Cold Fusion Institute is established in Utah.
October 16–18	A workshop on Anomalous Effects in Deuterated Materials, sponsored by the National Science Foundation and the Electric Power Research Institute, presents a relatively positive report on cold fusion and suggests that further research is justified.
November 12	The final report of the Department of Energy remains skeptical about the commercial possibilities inherent in cold fusion.
December	Positive results on cold fusion are reported from the Bhabha Atomic Research Center in India and from groups in Japan.

1990

January 1	Pons begins a major series of new experiments involving sixty-four cold-fusion cells.
February 1	Fritz G. Will takes over as head of the National Cold Fusion Institute.

March 28–31   The first annual Cold Fusion Conference is held in Salt Lake City.

March 29   A paper in *Nature* announces that independent measurements taken on Pons cells fail to find evidence of a nuclear reaction.

A major public blow to cold fusion came with the Department of Energy's interim report, released on July 12, 1989. The DOE had established a panel of experts who were to review the available data on cold fusion and visit the major laboratories. Their interim report was far from encouraging. There was no convincing evidence, they said, that cold fusion would ever provide energy on a commercial basis and no justification for establishing special research institutes. While this was not quite a total rejection of the scientific possibility of room-temperature nuclear fusion, as far as the general public was concerned, it certainly seemed to pour cold water on Pons and Fleischmann.

The DOE's interim report came hard on the heels of the decision by a major British laboratory to abandon its investigation of cold fusion. The nuclear research laboratory at Harwell is the birthplace of Britain's nuclear industry and the center of considerable expertise in nuclear technology. All these resources were brought to bear upon the subject of cold fusion. Indeed, the investigation itself reportedly cost 320,000 pounds (approximately $544,000) and used 4 million pounds' ($6.8 million) worth of equipment. Of course, no amount of costly equipment ever outweighs a carefully designed experiment and a good experimenter. Nevertheless, despite all the equipment and effort, including a total of fifty-six neutron detectors, the Harwell group could find no evidence of nuclear fusion.

Negative results were also reported from national laboratories in Germany, Switzerland, and Belgium. But the Harwell press conference, held on June 15 to announce the abandonment of the experiment, hit especially hard because of Martin Fleischmann's earlier involvement with the group.

On the positive side, as shown in Figure A-2, results continued to come in that suggested cold fusion is a real effect. In particular, Texas A&M, an early supporter of cold fusion, is now reporting the detection of large amounts of tritium, a by-product of nuclear fusion.

Finally, in August 1989, the National Cold Fusion Institute was established by the University of Utah and initially funded by the state. Hugo Rossi took over as director of this institute, which was located in the University of Utah's Research Park and, while separate from the university, was staffed mainly by University of Utah faculty with some others from Brigham Young University as well as visiting researchers.

Figure A–2

SOME LABORATORIES REPORTING
POSITIVE EXPERIMENTAL RESULTS

Bhabha Atomic Research Center
Brigham Young University
Brookhaven National Laboratory
Case Western Reserve University
University of Florida
University of Hokkaido
Los Alamos National Laboratory
University of Minnesota
Moscow State University
Nagoya University
University of Nebraska
Oak Ridge National Laboratory
Oregon State University
Osaka University
University of Rome
Stanford Research Institute
Stanford University
Texas A&M University (two groups)
Upsala University
University of Utah (two groups)

Source: Data taken from personal communication with James Brophy, Vice President of Research, the University of Utah, Salt Lake City.

The aim of the National Cold Fusion Institute is not only to continue scientific research into the phenomenon of cold fusion but also to develop its practical applications. The institute is divided into basic and applied research divisions with research teams in physics, electrochemistry, metallurgy, and engineering. Pons and Fleischmann, of course, formed the backbone of the electrochemistry group and in January began an elaborate experiment involving a total of sixty-four cold-fusion cells. Not only would more sophisticated heat measurements be made on these cells, but the two scientists would also carry out an analysis for tritium gas.

The institute's metallurgy group would focus on analyzing electrodes that had been used in fusion cells. The aim of this analysis is to uncover the essential features that would make cold fusion reproducible.

The original goal of the engineering group was to investigate the implications of a high output of energy coming from fusion cells. But, as always, the main drawback of cold-fusion research has been its lack of reproducibility—heat just cannot be produced on demand. To date, therefore, the group has focused on designing and building better experiments and apparatus for measuring heat output.

The task of the physics group is to look for evidence of nuclear reactions, such as neutrons, as well as x-rays and gamma rays, which should be produced as nuclear particles race through the palladium. It was these experiments that would later give rise to controversy between Stanley Pons and members of the university's physics department.

In November 1989 Hugo Rossi resigned as director of the National Cold Fusion Institute. He was later replaced, effective February 1, 1990, by Fritz G. Will, a research scientist from General Electric's research laboratories at Schenectady, New York. Will is experienced not only as a scientist but also in directing industrial research programs. As a researcher, he had experience in the study of electrode surfaces—which appear to be a key issue in obtaining reproducibility of cold fusion. He had also been involved in the cold-fusion collaboration between GE and the University of Utah. A year after the announcement of cold fusion by Pons and Fleischmann, Will was saying, "The multitude of results obtained by so many different groups can no longer be explained away as experimental artifacts." In his opinion, cold fusion is real!

November also saw the final report of the Department of Energy, which remained skeptical about the future of cold fusion: "The experimental results on excess energy from calorimetric cells reported to date do not present convincing evidence that useful sources of energy will result from the phenomena attributed to cold fusion." However, the DOE's position was softened somewhat, for the report did acknowledge that there were unexplained results that needed looking at, so that further research in this field was justified.

A month earlier, a workshop jointly sponsored by the National Science Foundation and the Electric Power Research Institute (which had been funding a number of cold-fusion experiments) was more positive. The meeting, held in Washington on October 16–18, concluded, "The results cannot be explained as a result of artifacts, equipment error or human errors." The workshop's participants were, however, concerned with the problem of reproducing results and with the lack of agreement between different researchers. They concluded, "Given the potential significance of the problem, further research is definitely required to improve the reproducibility of the effects and to unravel the mystery of the observations."

Cold fusion also received the support of a powerful scientific figure from the nuclear past, Edward Teller, who has been called the father of the U.S. hydrogen bomb. Teller, who attended the Washington workshop, pointed out that, according to established nuclear theory, the cold fusion simply should not take place. So if fusion is really occurring, then it has to be caused by some new process. One idea he proposed was that it is catalyzed by a new and as yet unknown elementary particle. Teller concluded, "It is recommended in recognition of the high-class work that yielded surprising results that the effort be supported in order to obtain clarification." He also suggested replacing palladium in the cells with an isotope of uranium and using beryllium instead of deuterium.

There had also been some speculation concerning the extent to which cold fusion was connected with the earlier work on muon-catalyzed fusion that had been pioneered by Steven Jones. Jones had shown that replacing one of the electrons that orbits a deuterium nucleus with a much heavier elementary particle called a muon has the effect of pulling two deuterium nuclei closer together. In this way, the nuclear fusion between them can be enhanced. Despite this enhancement, however, fusion could never go fast enough to produce measurable heat. But scientists began to wonder whether muons could have an effect within the deuterium-saturated palladium. However, researchers at MIT found no effect when they conducted experiments in which a muon beam was directed into palladium metal.

The major occasion for an overview of cold-fusion research came with the first annual Cold Fusion Conference. This was held on March 28–31, just over a year after the first press announcement, at the University Park Hotel close to the campus of the University of Utah and was sponsored by the National Cold Fusion Institute. The some 230 scientists who had signed up to attend could, for the first time, meet together in a major group and discuss their research. Some critics of the meeting felt that most of the invited speakers were cold-fusion "believers," with insufficient weight given to scientists who are critical of the idea.

The conference began in a spirit of great controversy. Press stories reported that scientists who had actually worked in Pons's own laboratory denied the existence of cold fusion. This gave rise to considerable recriminations and even to the rumor of a pending lawsuit between Pons and another staff member at the university.

One of the early puzzles concerning cold fusion had been the considerable success of Pons and Fleischmann compared to other researchers. Therefore, what better confirmation of the whole phenomenon than to measure nuclear fusion as it happened in Pons's own cells? This is what Mike Salamon and his colleagues from the University of Utah's department of physics planned to do. They set up their detectors in Stanley Pons's laboratory, underneath the table on which several cold-fusion cells were working. The apparatus was capable of registering the passage of neutrons, gamma rays, electrons, and protons. Nevertheless, over the period of their experiment, Salamon and his group could find no evidence of any nuclear fusion taking place. "We did not see a peep," said Salamon. "There was not an iota, not a sniff, of conventional fusion occurring."

This looked like damning evidence for Pons. How could he claim evidence of nuclear fusion when an independent group could find nothing. Pons, however, argued that during the period the physicists had chosen to set up their apparatus, the cells were working at "barely detectable levels," so he was not surprised that no fusion was detected. Indeed, he had been asked to review the paper for *Nature* but had refused, being too closely connected with the experiment. However, he had pointed out some inadequacies in the experiment, and these criticisms appeared to have been ignored by *Nature*. But, said the cold-fusion skeptics, this was the same old story over again. Why can only Stanley Pons produce results? Why did the reaction never work when anyone else was looking?

Then the controversy escalated. Pons went on to claim that

during a two-hour period, following a lightning strike when the nuclear detection instruments had been knocked out, one of his cells had released excess heat. But Salamon had a scientific trick up his sleeve. Although the electronics in his detectors had been turned off by the lightning strike, the evidence of any nuclear particle would still be present in the detectors themselves. Indeed, it should be possible to look for evidence of such particles several hours later.

The physicists went back and checked their data, only to find nothing—no evidence of nuclear decay during the very period when the cell was supposed to be active. As far as Ronald Parker, director of MIT's plasma fusion center was concerned, "It's another nail in the coffin." This fresh negative evidence certainly seemed damning, particularly as it was announced on the eve of the Cold Fusion Conference.

The timing of Salamon's paper to appear in *Nature* on March 29, right in the middle of the first annual Cold Fusion Conference, angered Pons and Fleischmann. (Stories in the press began to appear a day or two earlier.) They maintained it was yet another example of *Nature*'s policy of publishing negative papers on cold fusion. And as to the timing, "it is shameful and incredible that *Nature* had embargoed the publication of this paper until today with the obvious intent of trying to discredit the cold-fusion meeting in Utah," Pons said. John Maddox, the editor of *Nature*, was becoming something of a villain to the staunch supporters of cold fusion.

Salamon's paper notwithstanding, the general feeling of the conference was that something is definitely going on, possibly a nuclear reaction. But lack of reproducibility remains a major problem. As to the promise of free and abundant energy, "we're not in a position to say it's practical. It doesn't mean we won't be able to some day, but certainly we can't do it today," said Charles Scott of Oak Ridge National Laboratory.

The conference opened with a talk by Stanley Pons and ended with one by Martin Fleischmann, which earned him a standing ovation. Pons's talk was eagerly awaited, for people expected that at last the wall of silence around the experiments would be broken. Pons attempted to address the criticisms that had been used against his claims for cold fusion and gave details of how he and his colleagues had calculated the amount of heat given out by their cells. "We stand exactly where we did a year ago," he said, "with a great number of additional experiments and techniques to be reported. Our results are almost identically the same as then. We are more convinced now. If we were 99 percent sure [then], we are 100 percent sure [now]."

A twenty-five-page paper by Pons and Fleischmann was distributed during the meeting. Published in the summer of 1990 by the *Journal of Fusion Technology*, it was titled "Our Calorimetric Measurements of the Pd/D System: Fact and Fiction." (A calorimeter is a piece of equipment used to measure heat.) Other papers by Pons and Fleischmann were due to appear in print.

Their conclusion was that, through careful experimentation and by accounting for all the various effects involved, one must take as genuine the heat production in their cells. It is not simply a matter of heat being occasionally emitted from a charged cell, but of more energy actually being given out over an extended period than is ever pumped into the cell (for example, by electrical charging of the cell). In addition, there are "bursts" or excursions of heat in which the output of energy from the cell suddenly increases for a matter of hours—in one case for as long as 500 hours. The overall amount of heat given out is 100 to 1,000 times larger than could ever be created in any chemical reaction.

From these observations, Pons and Fleischmann concluded that cold fusion is indeed a nuclear process. This was confirmed by the absence of any heat when light water replaces heavy water in the cells. Chemically the two forms of water are virtually identical, so the absence of heat when light water is used points away from a chemical reaction to something nuclear.

Other speakers at the meeting reported additional confirmations of cold fusion, not only from national laboratories in the United States but also from major laboratories in India and Japan. The conclusion was that while heat is being detected by a number of groups, reproducibility remains a major problem. It is also difficult to find any direct evidence of nuclear fusion that is associated directly with a burst of heat. The most promising evidence of a nuclear reaction comes from the detection of tritium, a heavy, radioactive isotope of hydrogen, large amounts of which have been found in a number of fusion cells.

David Worledge, of the Electric Power Research Institute, while trying to keep a balanced view of the whole field, now says that the evidence is coming down in favor of cold fusion as a nuclear process. A variety of objections could be made to some of the earlier experiments that claimed to see cold fusion, but now, thanks to careful redesigning of experiments, it appears to Worledge that "most of the holes have been plugged. It's beginning to get where you have to contrive miracles to explain away the results rather than accept them." For Worledge, the detection of tritium has been a major factor, although the difficulty of getting heat production and other effects time and time again, and in an exactly reproducible way, has been one of the reasons why funding has not been more readily available.

Although a variety of new results on cold fusion had been presented over the winter months, the first annual Cold Fusion Conference provided a forum for consolidating the various data on cold fusion. The national laboratories at Los Alamos and Oak Ridge, for example, claimed to have detected heat and the products of a nuclear reaction in their cells, while Sandia National Laboratory had not. The Los Alamos group ran nine cells and detected excess tritium in seven of them—in one case, an eightyfold increase over several days of investigations.

Of particular interest have been results coming from outside the United States. The Japanese, in particular, have been actively looking at cold fusion. In addition to a number of universities, institutions, and private companies, the Japanese Fusion Institute, dedicated to the development of high-temperature fusion, has been researching cold fusion and has some positive results.

In India the Bhabha Atomic Research Center also has been doing careful work on cold fusion. Six different groups have been working in the center. India would be well served by a supply of energy that need not be geared to expensive technology. The Bhabha group has had a long experience in working with Canadian-designed nuclear reactors (fission reactors) like CANDU that use heavy water. During the normal operation of such a reactor, tritium builds up as a by-product in the heavy water, and it is therefore important to measure tritium levels. As a consequence, the Bhabha group had considerable experience in handling and monitoring tritium, so its results on cold fusion were looked on with some anticipation.

The group showed that in some fusion cells, the tritium content was twenty thousand times higher than in normal heavy water. The

Bhabha results are a significant indication that some sort of nuclear reaction is taking place. However, although conventional theory suggests that neutrons should be produced along with tritium, these neutrons are simply not being detected, or at least only in very low levels. The disparity between low neutron levels and high tritium levels remains a puzzle; on the other hand, the combination of heat and tritium is very persuasive. "The big question is what kind of fusion is going on," said Edmund Storms of Los Alamos National Laboratory, who suggests more than one nuclear reaction could be taking place in the cell.

Of course, there is no end of cold-fusion theories. The Nobel laureate Julian Schwinger presented his own ideas, as did Guilano Preparata of Milan, who with Emilio Del Giudice and T. Bressani proposed what Preparata called a "superradiant field" to explain the results. Essentially, cooperative interactions between the electrically charged particles in the palladium lattice and the electromagnetic field conspire to reduce the repulsion between deuterium nuclei and hold them together close enough and long enough to fuse—a theory that is endorsed by Martin Fleischmann.

Preparata and Del Giudice are strong and persuasive figures. Preparata has said, "I think we should have the courage to say, if we feel so, that it is cold fusion. It has been proven to the extent that these phenomena are real, although they are not controllable. . . . The world is well advised to take it seriously."

As to heat, this was still being detected not only by Pons and Fleischmann, but by groups like that of John Bockris of Texas A&M, who in one case ran a group of cells from November 2 to February 15 with no results until one of the cells suddenly began to give off heat. Charles Scott from Oak Ridge claimed that his cells were giving off 5 to 10 percent more energy than was being put in.

At the end of the meeting, a number of participants were unconvinced, but others like Richard Petrasso of MIT admitted, "I'm still skeptical, but I think there are some exciting things that need to be explained." The general verdict was that something curious was going on, but for Stanley Pons, the conclusion was even stronger: "You can't deny there's something here. It may be a huge thing, it may not, but you have to look at it."

The meeting was also not without its humorous notes, as when Steven Jones announced a joint experiment to be carried out between Brigham Young University and the University of Utah. This collaboration, he said, deserves the Nobel Peace Prize!

Even after the first annual International Cold Fusion Conference had ended, the topic continued to attract controversy. An article in the July 1990 *Science* again raised the question of possible fraud in some of the Texas A&M experiments.

Groups led by John Bockris and John Appleby at Texas A&M University are among the strongest supporters of the Pons and Fleischmann brand of cold fusion. Bockris, in particular, had discovered a dramatic increase in the concentration of tritium when his fusion cells were working. But to some critics this concentration was just too good to be true and the possibility of fraud was raised. Even John Appleby is reported to have asked Bockris, "Look, concerning this tritium—are you sure that someone hasn't been spiking your cells?" The implication being that someone may have added a few drops of tritiated water to the experimental cells (a small bottle of tritiated water was kept in the laboratory).

The accusation of scientific fraud is extremely serious. One suggestion to resolve the issue was to repeat the same experiment in another laboratory that would be locked at all times and allow only limited access to staff. Bockris was prepared to discount the idea of fraud since, as he wrote in the *Journal of Electroanalytical Chemistry*, "Interference with the experiments is considered improbable because of positive results from the Cyclotron Institute to which entrance is prohibited except by the usual personnel at the Institute." However it appears that the building was not guarded at night or on the weekends and a number of researchers had access to keys.

In a related development, Kevin Wolf, another researcher at Texas A&M who had been involved in positive tritium detection, now believes those results were caused by contamination in the cold fusion cells. (Several of the labortories that detected tritium had obtained the palladium used in their cold fusion cells from the same supplier, Hoover and Strong, Inc., of Richmond, Virginia.) Concerned about impurities, Wolf first dissolved several samples of palladium in acid and analyzed the resulting solution for tritium. Sure enough it was found not only in the palladium from cold fusion cells but also in the pure palladium and in electrodes from blank cells. The tritium was therefore present as a contaminant in the metal from the beginning.

But the amount of tritium contamination did not explain the high concentrations seen by Bockris. Wolf went on to analyze the electrolytes from several of Bockris's cells and discovered that they contained a large amount of light water—ordinary water. This would certainly be consistent with the idea that the cells had been spiked with tritiated water, which contains a large percentage of light water. However, the same sort of contamination would also be present if the cells had been exposed to the air.

Wolf's findings certainly do not demolish Bockris's results but they do cast considerable doubt upon those results. As Kevin Wolf puts it "The proper conclusion is that things [in Bockris's laboratory] were so uncontrolled and sloppy [that] those studies don't mean anything."

So what is the verdict on cold fusion? Is it fact or fiction? Reality or fakery? The conclusions, as of July 1990, are as follows.

HEAT

A number of laboratories have detected "bursts" or excursions of heat from cold-fusion cells. A cell can run for many days, even for weeks, before suddenly increasing its output of heat. While these heat excursions, which can be quite large and last for tens and even hundreds of hours, are difficult to reproduce, their existence is reasonably well established.

In addition to occasional bursts of heat, there is a variety of claims that continuous heat at a lower level is being produced by working cells. This result is more controversial, since it relies upon being able to measure both the total amount of energy that is supplied to the cells—for example, during the long period of charging and in normal running—and the amount of heat that is being given out by the cell. If cold fusion truly works, then in the long run, more heat should be given out than is put into the cell. But such measurements are very difficult to carry out; a variety of potential errors and other factors require many compensations and assumptions, all of which have been questioned by cold-fusion critics. However, the design of apparatus and the experiments is improving, and it should eventually be possible to establish definitive results one way or the other.

With a net heat output still not fully established, it is possible that sudden bursts of heat could have a nonnuclear origin. For example, the part of the energy that is pumped into the cell during its normal operation is perhaps being stored in some way to be released later as a burst of heat. In this sense, a cold-fusion cell would be acting like a battery, alternately being charged with energy and then releasing it. A major factor, therefore, will be to establish convincing evidence of an *overall* energy output from a fusion cell.

Excess heat has been claimed at Oak Ridge, Texas A&M University, and several other centers. Pons and Fleischmann claim to have measured such net heat output and point out that it is 100 to 1,000 times higher than can be generated by any chemical process. If it is true that excess heat of such a magnitude can be consistantly produced over hundreds of hours, then some sort of nuclear process must clearly be involved. Powerful independent confirmation of this would come from the detection of the by-products of a nuclear reaction. There is some evidence that this has already been established.

NEUTRONS

In any conventional theory of cold fusion, sudden excursions of heat would be accompanied by intense bursts of neutrons, but these have never been observed. Theoreticians have therefore turned to a variety of theories involving nuclear processes that produce either very few neutrons or no neutrons at all.

In their first paper, Pons and Fleischmann claimed to have observed some neutrons. Jones for his part also observed neutrons, but at a much lower level. There were even claims that bursts of neutrons are associated with bursts of heat. However, many technical difficulties arise in interpreting the results of neutron measurements. While a number of groups, such as Oak Ridge National Laboratory and groups at Hokkaido, Nagoya, and Osaka universities, claim to have detected neutrons, others see no evidence of neutrons being emitted from working cells. The evidence for neutrons therefore still lies in the balance.

TRITIUM

A variety of exotic nuclear processes have been proposed to account for the nuclear fusion of deuterium without the emission of neutrons.

But conventional physics and most of the exotic fusion theories agree that tritium, a radioactive isotope of hydrogen, should be produced. Therefore, the detection of tritium in fusion cells would be incontrovertible evidence that a nuclear process has taken place. But, again, such results need careful analysis to ensure, for example, that the sample is not contaminated in any way, or that tritium was not already present in the cell.

The presence of radioactive tritium has been established by a number of laboratories and is the most compelling argument that some sort of nuclear reaction is taking place in a cold-fusion cell. Strong results on tritium detection come from the Bhabha Atomic Research Center as well as from Los Alamos and Oak Ridge National Laboratories. However, allegations of contamination and of "spiking" fusion cells with tritiated water have cast doubt on some results, from Texas A&M in particular.

REPRODUCIBILITY

As mentioned, one of the major problems in the cold-fusion story is that of reproducibility. One researcher can make cold-fusion cells function; another working with the same recipe never has any success. Not even cold-fusion believers can always get a cell to work, and "heat bursts" cannot be produced to order.

Not only is the phenomenon itself not reproducible, but the reports of what is seen differ from laboratory to laboratory. Some labs observe high levels of tritium; others see nothing, even during heat bursts.

Cold fusion remains elusive and confusing. Just what is going on in those test tubes? Is some new and exotic nuclear reaction occurring? Or perhaps a number of quite different processes are taking place, some of which are seen by one researcher and not by others.

A year after the Pons and Fleischmann announcement, we are no closer to a definitive account of cold fusion. Nature appears to be playing some of the cards close to the chest. Only time will tell what is really going on when an electrical current is passed through a test tube containing heavy water and a palladium electrode.

Appendix 1
Radiation and Radioactivity

It is easy to become confused by the difference between radiation and radioactivity. Radioactivity is the property of some substances—like radium, uranium, and plutonium—to give off nuclear radiation.

RADIATION

There are many types of radiation, such as heat and sound, as well as the various forms of electromagnetic radiation, such as radio waves, infrared, visible light, ultraviolet light, x-rays, and gamma rays.

Nuclear radiation that is emitted by radioactive substances, nuclear power plants, and fusion reactions can be subdivided into the following forms:

- Alpha particles—The nuclei of helium atoms, alpha particles, consist of two neutrons and two protons that are bound together. Alpha particles are given off by some radioactive substances such as plutonium and are easily stopped by a few thicknesses of paper.

- Beta rays—These are fast electrons emitted from radioactive substances like tritium. Beta rays can be stopped by a thick metal foil.

- Neutrons—These particles are given off in the radioactive decay of some uranium isotopes and by the fusion of two deuterium atoms to produce helium 3. Since neutrons have no electrical charge, it is difficult to "stop" them, and neutron radiation is very penetrating. Concrete shielding several feet thick is required to shield the neutrons' radiation from a nuclear reactor. In addition, a bombardment of neutrons will make certain substances radioactive.

- Gamma rays—These are the highly energetic portion of the electromagnetic spectrum, which begins at the x-ray end.

The radiation from a dental x-ray machine ceases as soon as the device is turned off. There is no residual radiation lurking around the office any more than residual light remains in a room after an incandescent bulb is switched off.

Exactly the same thing happens with a nuclear reactor; as soon as the reactor is shut down, the fission reaction ends, and radiation from this process ceases. However, the by-products of the fission process are constantly accumulating in the fuel rods. These consist of highly radioactive substances that are constantly emitting radiation. Hence, even when a nuclear power station is shut down, nuclear radiation continues to be emitted from the radioactive isotopes that have been created in the fuel rods.

Gamma rays and neutrons are both produced in a tokamak when it is running. As soon as the tokamak is turned off, this radiation ceases. However, the neutrons produced in the daily operation of a tokamak are bombarding the nuclei of the materials used to construct the device and the shielding around it. These neutrons make this material radioactive so that even when a tokamak is shut down, it may still be giving off nuclear radiation.

Radioactive contamination occurs when a radioactive substance, in the form of a fine dust or gas, is present in a building. These radioactive substances continue to give off nuclear radiation and must be cleaned up and removed.

The nuclear radiation from some radioactive isotopes may be weak and of a short range but may become more hazardous if the isotopes are ingested. Some radioactive substances enter the food chain and can be absorbed by the human body. Radioactive iodine, for example, will accumulate in the thyroid, while radioactive strontium is deposited in the bones.

HEALTH EFFECTS

Doctors distinguish between what they call ionizing and nonionizing radiation. Ionizing radiation, like x-rays and nuclear radiation, has enough energy to knock electrons from atoms and molecules. This radiation could damage the DNA within a human cell, resulting in cancer, birth defects, and genetic defects in future generations.

Nonionizing radiation, like sunlight, ultrasound, or the radio waves from a magnetic resonance imaging (MRI) scanner, cannot break up molecules in the cell in this way. However, other adverse biological effects cannot be ruled out.

Appendix 2
The Utah Fusion Cell

The Utah fusion cell contains 99.5 percent heavy water with 0.5 percent light water. (Also, 0.1 M of LiOD was added to the cell.)

Electrical current for a given area of electrode surface: Highest used was 512 ma/cm^2.

Neutron flux: 4,000/sec for the 0.4 cm diameter electrode.

Heat output: Up to 1,224 percent of break-even value—that is, up to 21 W/cm^3 of the electrode.

Projected heat output: Using mixtures of heavy water containing deuterium and tritium, the excess heat could reach 1 million times that put into the cell—that is, 10 kW/cm^3!

Configurations:

- *A*—Negative electrode: a palladium sheet. Positive electrode: platinum sheet.

- *B*—Negative electrode: 10 cm palladium rods with diameters of 1, 2, and 4 mm. Positive electrode: platinum wire wound on a cage of glass rods.

- *C*—Negative electrode: a palladium cube 1 cm × 1 cm × 1 cm. **Warning! Ignition!** In this case, the electrode heated to its melting point (1,554° C) vaporized and destroyed part of the cupboard housing the experiment.

Appendix 3
The Brigham Young Cell

Contents: Heavy water laced with the following salts: $FeSO_4$, $NiCl_2$, $PdCl_2$, $CaCO_3$, Li_2SO_4, Na_2SO_4, $CAH_4(PO_4)_2$, $TiOSO_4$, Na_2SO_4, $AuCN$.

Electrical current: 10–50 ma at 3–25.

Heat output: None detected.

Fusion rate: 10^{-23} fusions/pair of deuterons/sec.

Configurations: 4–8 cells used simultaneously, each consisting of a small jar 4 cm high and 4 cm in diameter containing 20 ml of solution. Negative electrode: 1 g "fused" titanium pellets; 0.025 mm thick palladium foils; 5 g mossy palladium.

Index